THE GENERAL THEORY OF EMPLOYMENT, INTEREST AND MONEY

THE GENERAL THEORY OF EMPLOYMENT, INTEREST AND MONEY

JOHN MAYNARD KEYNES

for the Royal Economic Society

First edition published 1936
This edition published 2007 by
PALGRAVE MACMILLAN
Houndmills, Basingstoke, Hampshire RG21 6XS and
175 Fifth Avenue, New York, N. Y. 10010
Companies and representatives throughout the world

PALGRAVE MACMILLAN is the global academic imprint of the Palgrave
Macmillan division of St. Martin's Press, LLC and of Palgrave Macmillan Ltd.
Macmillan® is a registered trademark in the United States, United Kingdom
and other countries. Palgrave is a registered trademark in the European
Union and other countries.

ISBN-13: 978–0230–00476–4   paperback
ISBN-10: 0–230–00476–8   paperback

This book is printed on paper suitable for recycling and made from fully
managed and sustained forest sources. Logging, pulping and manufacturing
processes are expected to conform to the environmental regulations of the
country of origin.

A catalogue record for this book is available from the British Library.

A catalogue record for this book is available from the Library of Congress.

10   9   8   7   6
16   15   14   13   12   11   10   09

Printed and bound in Great Britain by
CPI Antony Rowe, Chippenham and Eastbourne

Contents

Book V Money-wages and Prices

Book VI Short Notes Suggested by the General Theory

General Introduction

This new standard edition of *The Collected Writings of John Maynard Keynes* forms the memorial to him of the Royal Economic Society. He devoted a very large share of his busy life to the Society. In 1911, at the age of twenty-eight, he became editor of the *Economic Journal* in succession to Edgeworth; two years later he was made secretary as well. He held these offices without intermittence until almost the end of his life. Edgeworth, it is true, returned to help him with the editorship from 1919 to 1925; MacGregor took Edgeworth's place until 1934, when Austin Robinson succeeded him and continued to assist Keynes down to 1945. But through all these years Keynes himself carried the major responsibility and made the principal decisions about the articles that were to appear in the *Economic Journal*, without any break save for one or two issues when he was seriously ill in 1937. It was only a few months before his death at Easter 1946 that he was elected president and handed over his editorship to Roy Harrod and the secretaryship to Austin Robinson.

In his dual capacity of editor and secretary Keynes played a major part in framing the policies of the Royal Economic Society. It was very largely due to him that some of the major publishing activities of the Society—Sraffa's edition of Ricardo, Stark's edition of the economic writings of Bentham, and Guillebaud's edition of Marshall, as well as a number of earlier publications in the 1930s—were initiated.

When Keynes died in 1946 it was natural that the Royal Economic Society should wish to commemorate him. It was perhaps equally natural that the Society chose to commemorate him by producing an edition of his collected works. Keynes himself had always taken a joy in fine printing, and the Society, with the help of Messrs Macmillan as publishers and the Cambridge University Press as printers, has been anxious to give Keynes's writings a permanent form that is wholly worthy of him.

The present edition will publish as much as is possible of his work in the field of economics. It will not include any private and personal correspondence or publish letters in the possession of his family. The edition is concerned, that is to say, with Keynes as an economist.

Keynes's writings fall into five broad categories. First, there are the books which he wrote and published as books. Second, there are collections of articles and pamphlets which he himself made during his lifetime (*Essays in Persuasion* and *Essays in Biography*). Third, there is a very considerable volume of published but uncollected writings—articles written for newspapers, letters to newspapers, articles in journals that have not been included in his two volumes of collections, and various pamphlets. Fourth, there are a few hitherto unpublished writings. Fifth, there is correspondence with economists and concerned with economics or public affairs.

This series will attempt to publish a complete record of Keynes's serious writing as an economist. It is the intention to publish almost completely the whole of the first four categories listed above. The only exceptions are a few syndicated articles where Keynes wrote almost the same material for publication in different newspapers or in different countries, with minor and unimportant variations. In these cases, this series will publish one only of the variations, choosing the most interesting.

The publication of Keynes's economic correspondence must inevitably be selective. In the day of the typewriter and the filing cabinet and particularly in the case of so active and busy a man, to publish every scrap of paper that he may have dictated about some unimportant or ephemeral matter is impossible. We are aiming to collect and publish as much as possible, however, of the correspondence in which Keynes developed his own ideas in argument with his fellow economists, as well as the more significant correspondence at times when Keynes was in the middle of public affairs.

Apart from his published books, the main sources available to those preparing this series have been two. First, Keynes in his will made Richard Kahn his executor and responsible for his economic papers. They have been placed in the Marshall Library of the University of Cambridge and have been available for this edition. Until 1914 Keynes did not have a secretary and his earliest papers are in the main limited to drafts of important letters that he made in his own handwriting and retained. At that stage most of the correspondence that we possess is represented by what he received rather than by what he wrote. During the years 1914–18 and 1940–46 Keynes was serving in the Treasury. With the opening of official records, many of the papers that he wrote have become available. From 1919 onwards, throughout the rest of his life, Keynes had the help of a secretary—for many years Mrs Stevens. Thus for the last twenty-five years of his working life we have in most cases the carbon copies of his own letters as well as the originals of the letters that he received.

There were, of course, occasions during this period on which Keynes wrote himself in his own handwriting. In some of these cases, with the help of his correspondents, we have been able to collect the whole of both sides of some important interchanges and we have been anxious, in justice to both correspondents, to see that both sides of the correspondence are published in full.

The second main source of information has been a group of scrapbooks kept over a very long period of years by Keynes's mother, Florence Keynes, wife of Neville Keynes. From 1919 onwards these scrapbooks contain almost the whole of Maynard Keynes's more ephemeral writing, his letters to newspapers and a great deal of material which enables one to see not only what he wrote, but the reaction of others to his writing. Without these very carefully kept scrapbooks the task of any editor or biographer of Keynes would have been immensely more difficult.

The plan of the edition, as at present intended, is this. It will total twenty-five volumes. Of these, the first eight will be Keynes's published books from *Indian Currency and Finance*, in 1913, to the *General Theory* in 1936, with the addition of his *Treatise on Probability*. There will next follow, as vols. IX and X, *Essays in Persuasion* and *Essays in Biography*, representing Keynes's own collections of articles. *Essays in Persuasion* will differ from the original printing in two respects: it will contain the full texts of the articles or pamphlets included in it and not (as in the original printing) abbreviated versions of these articles, and it will have added one or two later articles which are of exactly the same character as those included by Keynes in his original collection. In the case of *Essays in Biography*, we shall add one or two other biographical studies that Keynes wrote later than 1933.

There will follow four volumes, XI to XIV, of economic articles and correspondence, and one volume of social, political and literary writings. We shall include in these volumes such part of Keynes's economic correspondence as is closely associated with the articles that are printed in them.

The further nine volumes, as we estimate at present, will deal with Keynes's *Activities* during the years from the beginning of his public life in 1905 until his death. In each of the periods into which we propose to divide this material, the volume concerned will publish his more ephemeral writings, all of it hitherto uncollected, his correspondence relating to these activities, and such other material and correspondence as is necessary to the understanding of Keynes's activities. These volumes are being edited by Elizabeth Johnson and Donald Moggridge, and it is their task to trace and interpret Keynes's activities sufficiently to make the material fully intelligible to a later generation. Until this work has progressed further, it is not possible to say with exactitude whether this material will be distributed, as we now think, over nine volumes, or whether it will need to be spread over a further volume or volumes. There will be a final volume of bibliography and index.

Those responsible for this edition have been: Lord Kahn, both as Lord Keynes's executor and as a long and intimate friend of Lord Keynes, able to help in the interpreting of much that would otherwise be misunderstood; Sir Roy Harrod as the author of his biography; Austin Robinson as Keynes's co-editor on the *Economic Journal* and successor as secretary of the Royal Economic Society. The initial editorial tasks were carried by Elizabeth Johnson. More recently she has been joined in this responsibility by Donald Moggridge. They have been assisted at different times by Jane Thistlethwaite,

Mrs McDonald, who was originally responsible for the systematic ordering of the files of the Keynes papers; Judith Masterman, who for many years worked with Mrs Johnson on the papers; and more recently by Susan Wilsher, Margaret Butler, and Barbara Lowe.

Editorial Introduction

'I have been much pre-occupied with the causation, so to speak, of my progress of mind from the classical position to my present views,—with the order in which the problem developed in my mind. What some people think is an unnecessarily controversial tone is really due to the importance in my own mind of what I *used* to believe, and of the moments of transition which were for me personally moments of illumination.... You don't mention *effective demand* or, more precisely, the demand schedule for output as a whole, except in so far as it is implicit in the multiplier. To me the most extraordinary thing, regarded historically, is the complete disappearance of the theory of demand and supply for output as a whole, *i.e.* the theory of employment, *after* it had been for a quarter of a century the most discussed thing in economics. One of the most important transitions for me, after my *Treatise on Money* had been published, was suddenly realising this. It only came after I had enunciated to myself the psychological law that, when income increases, the gap between income and consumption will increase,—a conclusion of vast importance to my own thinking but not apparently, expressed just like that, to anyone else's. Then, appreciably later, came the notion of interest being the measure of liquidity preference, which became quite clear in my mind the moment I thought of it. And last of all, after an immense amount of muddling and many drafts, the proper definition of the marginal efficiency of capital linked up one thing with another.'

With these words, Keynes told R. F. Harrod in the summer of 1936 the development of his ideas towards the *General Theory*.[1]

The origins of the *General Theory* lie in Keynes's dissatisfaction with his *Treatise on Money* even at the time of publication,[2] in the prolonged international slump of the years after 1929, and in the stimulation that emanated from a 'circus' of young Cambridge economists who began meeting soon after the publication of the *Treatise* to discuss and dissect its two volumes. It was the discussions within this group, retailed to him by Richard Kahn, that provided the basis for the first transitional stage between the *Treatise* and the *General Theory*.[3]

This stage was soon followed by Keynes's explicit commitment to revise the theoretical foundations of the *Treatise*, which had only dealt incidentally in

---

[1] Keynes to R. F. Harrod, 30 August 1936. This letter appears in full in volume xiii.

[2] See, for example, his letter to his mother of 14 September, 1930 in JMK, vol. v, p. xv.

[3] A note on the circus and all surviving papers appear in volume xiii.

terms of movements in output. Thus in his preface for Japanese readers of the *Treatise*, dated April 1932, Keynes noted that rather than revise his *Treatise* he proposed 'to publish a short book of a purely theoretical character, extending and correcting the theoretical basis of my views as set forth in Books III and IV'.[4] This commitment became clearer in the autumn of 1932, when Keynes changed the title of his course of lectures from 'The Pure Theory of Money', their title since the autumn of 1929, to 'The Monetary Theory of Production', the title they were to have until 1934. These lectures were concerned with movements in output as a whole and had the beginnings of the concept of liquidity preference, although it was not until his lectures in the autumn of 1933 that it took the form used in the *General Theory*.

The first major published indications of the direction Keynes's thought was taking between the *Treatise* and the *General Theory* came in 1933 in the form of an essay 'The Monetary Theory of Production', a pamphlet 'The Means to Prosperity', an article 'The Multiplier' (which is included in the American edition of 'The Means to Prosperity'), and a biographical sketch of T. R. Malthus,[5] all of which are cast in terms of movements of output as a whole and reflected dissatisfaction with accepted theory.

Thus the major building blocks of the *General Theory* had been steadily accumulating ever since 1932. By the spring of 1934, in fact, all of them were in place, except for the idea of the marginal efficiency of capital, as is clear from drafts from that period and a working paper Keynes prepared during his visit to America in May and June.[6] It was only during the summer of 1934, however, that the final piece fell into place, and by the autumn Keynes was delivering his lectures, now entitled 'The General Theory of Employment' from proof sheets.

However, before publication there was to be another year of intense discussion and redrafting. Keynes circulated proofs of the book to R. F. Kahn, Joan Robinson, R. F. Harrod, D. H. Robertson and R. G. Hawtrey and took careful note of their comments and suggested improvements, stating clearly his points of disagreement when he did not adopt them.[7] Thus it was after almost five years of intense preparation that the book appeared in February 1936 at a price of 5 shillings to encourage a large sale among students.

---

[4] JMK vol. v, p. xxvii.
[5] 'The Monetary Theory of Production' appears in volume xiii, 'The Means to Prosperity' in volume ix (vi) i, 'Robert Malthus: The First of the Cambridge Economists' in volume x, Ch. 12.
[6] This paper will appear in volume xiii.
[7] This correspondence, keyed into the drafts and the final text of the *General Theory* appears in volume xiii.

After publication there was still further discussion and, occasionally, contro-versy. Keynes himself encouraged this discussion, for, as he put it,[8]

I am more attached to the comparatively simple fundamental that ideas underlie my theory than to the particular forms in which I have embodied them, and have no desire that the latter should be crystallized at the present stage of the debate. If the simple basic ideas can become familiar and acceptable, time and experience and the collaboration of a number of minds will discover the best way of expressing them.

In this spirit, Keynes entered into considerable correspondence with critics, expositors and extenders of his ideas.[9] Moreover, as the debate progressed, his own ideas were changing and by August 1936 he was writing to R. G. Hawtrey[10]

I may mention that I am thinking of producing in the course of the next year or so what might be called *footnotes* to my previous book, dealing with various criticisms and various points which want carrying further. Of course, in fact, the whole book wants re-writing and re-casting. But I am still not in a sufficiently changed state of mind as yet to be in the position to do that. On the other hand I can deal with specific points.

From this period, we have one draft table of contents to a book entitled 'Footnotes to *The General Theory of Employment and Money*', a title that echoes his first draft table of contents after *Economic Consequences of the Peace*.[11] He also used this title for the lectures he delivered in Cambridge in the spring of 1937, from which drafts of two lectures survive.[12] In fact he seems to have been making considerable progress towards stepping outside the *General Theory* by that time, for he told Joan Robinson in April 1937[13]

I am gradually getting myself into an outside position towards the book, and am feeling my way to new lines of exposition. Perhaps you will see what I have in mind in my forthcoming lectures.

Unfortunately, the proposed 'footnotes' never got beyond the lectures, for Keynes suffered a severe heart attack in the early summer of 1937 and was

---

[8] See Keynes's article 'The General Theory of Employment' (1937). This article appears in volume xiv.

[9] This correspondence appears in full in volume xiv.

[10] Keynes to R. G. Hawtrey, 31 August 1936. This letter appears in full in volume xiv.

[11] See *A Revision of the Treaty* (JMK, vol. iii) p. xiii.

[12] These appear in volume xiv.

[13] Keynes to Joan Robinson, 20 April 1937.

never able to work at anything near his old pace until war came in 1939—and then his energies were directed in other directions. How he would have revised the *General Theory* if he had remained in good health is impossible to guess. One can only be certain that he would have revised it.

Since its publication in Britain in February 1936, the *General Theory* has been published in the United States (originally from sheets printed in England) and translated into German, Japanese, French, Spanish, Czech, Italian, Serbo-Croat, Hindi, Finnish, Roumanian, Hungarian and Russian. The German, French and Japanese editions all carried special additional prefaces which followed the original English preface. These additional prefaces are printed below.

This edition follows the reprinted English first edition, which differs from the first English printing in that a correction to lines 23–5 of page 123 was moved from an erratum note following the index into the text. In Appendix 1 we have introduced a list of minor textual corrections from the reprinted first edition. In addition, in Appendices 2 and 3 we reprint Keynes's articles. 'Fluctuations in Net Investment in the United States' and 'Relative Movements of Real Wages and Output' which deal with errors on pages 103–4 and 9–10 respectively of the text itself. For further discussion of the *General Theory* and its genesis, the reader is referred to volumes XIII and XIV.

In printing this volume we have made it our first objective to follow as precisely as possible the pagination of the original edition. An immense literature of detailed criticism and analysis has grown up around the text of the original edition and we have been anxious that references in that literature should apply equally to this new edition. This has meant that we have been unable to follow precisely the standard typography of the other volumes in the series. The gain of ready reference, in our view, justifies this course. As in other volumes of the series, we have reduced the excessive capitalisation favoured by Keynes's original printers, but irritating to a modern eye.

Preface

This book is chiefly addressed to my fellow economists. I hope that it will be intelligible to others. But its main purpose is to deal with difficult questions of theory, and only in the second place with the applications of this theory to practice. For if orthodox economics is at fault, the error is to be found not in the superstructure, which has been erected with great care for logical consistency, but in a lack of clearness and of generality in the premisses. Thus I cannot achieve my object of persuading economists to re-examine critically certain of their basic assumptions except by a highly abstract argument and also by much controversy. I wish there could have been less of the latter. But I have thought it important, not only to explain my own point of view, but also to show in what respects it departs from the prevailing theory. Those, who are strongly wedded to what I shall call 'the classical theory', will fluctuate, I expect, between a belief that I am quite wrong and a belief that I am saying nothing new. It is for others to determine if either of these or the third alternative is right. My controversial passages are aimed at providing some material for an answer; and I must ask forgiveness if, in the pursuit of sharp distinctions, my controversy is itself too keen. I myself held with conviction for many years the theories which I now attack, and I am not, I think, ignorant of their strong points.

The matters at issue are of an importance which cannot be exaggerated. But, if my explanations are right, it is my fellow economists, not the general public, whom I must first convince. At this stage of the argument the general public, though welcome at the debate, are only eavesdroppers at an attempt by an economist to bring to an issue the deep divergences of opinion between fellow economists which have for the time being almost destroyed the practical influence of economic theory, and will, until they are resolved, continue to do so.

The relation between this book and my *Treatise on Money* [*JMK* vols. v and vi], which I published five years ago, is probably clearer to myself than it will be to others; and what in my own mind is a natural evolution in a line of thought which I have been pursuing for several years, may sometimes strike the reader as a confusing change of view. This difficulty is not made less by certain changes in terminology which I have felt compelled to make. These changes of language I have pointed out in the course of the following pages; but the general relationship between the two books can be expressed briefly as follows. When I began to write my *Treatise on Money* I was still moving along the traditional lines of regarding the influence of money as something so to speak separate from the general theory of supply and demand. When I finished it, I had made some progress towards pushing monetary theory back to becoming a theory of output as a whole. But my lack of emancipation from preconceived ideas showed itself in what now seems to me to be the outstanding fault of the theoretical parts of that work (namely, Books III and IV), that I failed to deal thoroughly with the effects of *changes* in the level of output. My so-called 'fundamental equations' were an instantaneous picture taken on the assumption of a given output. They attempted to show how, assuming the given output, forces could develop which involved a profit-disequilibrium, and thus required a change in the level of output. But the dynamic development, as distinct from the instantaneous picture, was left incomplete and extremely confused. This book, on the other hand, has evolved into what is primarily a study of the forces which determine changes in the scale of output and employment as a whole; and, whilst it is found that money enters into the economic scheme in an essential and peculiar manner, technical monetary detail falls into the background. A monetary economy, we shall find, is essentially one in which changing views about the future are capable of influencing the quantity of employment and not merely its direction. But our method of analysing the economic behaviour of the present under the influence of changing ideas about the future is one which depends on the interaction of supply and demand, and is in this way linked up with our fundamental theory of value. We are thus led to a more general theory, which includes the classical theory with which we are familiar, as a special case.

The writer of a book such as this, treading along unfamiliar paths, is extremely dependent on criticism and conversation if he is to avoid an undue proportion of mistakes. It is astonishing what foolish things one can temporarily believe if one thinks too long alone, particularly in economics (along with the other moral sciences), where it is often impossible to bring one's ideas to a conclusive test either formal or experimental. In this book, even more perhaps than in writing my *Treatise on Money*, I have depended on the constant advice and constructive criticism of Mr R. F. Kahn. There is a great deal in this book which would not have taken the shape it has except at his suggestion. I have also had much help from Mrs Joan Robinson, Mr R. G. Hawtrey and Mr R. F. Harrod, who have read the whole of the proof-sheets. The index has been compiled by Mr D. M. Bensusan-Butt of King's College, Cambridge.

The composition of this book has been for the author a long struggle of escape, and so must the reading of it be for most readers if the author's assault upon them is to be successful,—a struggle of escape from habitual modes of thought and expression. The ideas which are here expressed so laboriously are extremely simple and should be obvious. The difficulty lies, not in the new ideas, but in escaping from the old ones, which ramify, for those brought up as most of us have been, into every corner of our minds.

J. M. KEYNES

13 *December* 1935

Preface to the German Edition

Alfred Marshall, on whose *Principles of Economics* all contemporary English economists have been brought up, was at particular pains to emphasise the continuity of his thought with Ricardo's. His work largely consisted in grafting the marginal principle and the principle of substitution on to the Ricardian tradition; and his theory of output and consumption as a whole, as distinct from his theory of the production and distribution of a *given* output, was never separately expounded. Whether he himself felt the need of such a theory, I am not sure. But his immediate successors and followers have certainly dispensed with it and have not, apparently, felt the lack of it. It was in this atmosphere that I was brought up. I taught these doctrines myself and it is only within the last decade that I have been conscious of their insufficiency. In my own thought and development, therefore, this book represents a reaction, a transition away from the English classical (or orthodox) tradition. My emphasis upon this in the following pages and upon the points of my divergence from received doctrine has been regarded in some quarters in England as unduly controversial. But how can one brought up a Catholic in English economics, indeed a priest of that faith, avoid some controversial emphasis, when he first becomes a Protestant?

But I fancy that all this may impress German readers somewhat differently. The orthodox tradition, which ruled in nineteenth century England, never took so firm a hold of German thought. There have always existed important schools of economists in Germany who have strongly disputed the adequacy of the classical theory for the analysis of contemporary events. The Manchester School and Marxism both derive ultimately from Ricardo,—a conclusion which is only superficially surprising. But in Germany there has always existed a large section of opinion which has adhered neither to the one nor to the other.

It can scarcely be claimed, however, that this school of thought has erected a rival theoretical construction; or has even attempted to do so. It has been sceptical, realistic, content with historical and empirical methods and results, which discard formal analysis. The most important unorthodox discussion on theoretical lines was that of Wicksell. His books were available in German (as they were not, until lately, in English); indeed one of the most important of them was written in German. But his followers were chiefly Swedes and Austrians, the latter of whom combined his ideas with specifically Austrian theory so as to bring them in effect, back again towards the classical tradition. Thus Germany, quite contrary to her habit in most of the sciences, has been content for a whole century to do without any formal theory of economics which was predominant and generally accepted.

Perhaps, therefore, I may expect less resistance from German, than from English, readers in offering a theory of employment and output as a whole, which departs in important respects from the orthodox tradition. But can I hope to overcome Germany's economic agnosticism? Can I persuade German economists that methods of formal analysis have something important to contribute to the interpretation of contemporary events and to the moulding of contemporary policy? After all, it is German to like a theory. How hungry and thirsty German economists must feel after having lived all these years without one! Certainly, it is worth while for me to make the attempt. And if I can contribute some stray morsels towards the preparation by German economists of a full repast of theory designed to meet specifically German conditions, I shall be content. For I confess that much of the following book is illustrated and expounded mainly with reference to the conditions existing in the Anglo-Saxon countries.

Nevertheless the theory of output as a whole, which is what the following book purports to provide, is much more easily adapted to the conditions of a totalitarian state, than is the theory of the production and distribution of a given output produced under conditions of free competition and a large measure of laissez-faire. The theory of the psychological laws relating consumption and saving, the influence of loan expenditure on prices and real wages, the part played by the rate of interest—these remain as necessary ingredients in our scheme of thought.

I take this opportunity to acknowledge my indebtedness to the excellent work of my translator Herr Waeger (I hope his vocabulary at the end of this volume[1] may prove useful beyond its immediate purpose) and to my publishers, Messrs Duncker and Humblot, whose enterprise, from the days now sixteen years ago when they published my *Economic Consequences of the Peace*, has enabled me to maintain contact with German readers.

J. M. KEYNES

7 September 1936

---

[1] Not printed in this edition [Ed.].

Preface to the Japanese Edition

Alfred Marshall, on whose *Principles of Economics* all contemporary English economists have been brought up, was at particular pains to emphasise the continuity of his thought with Ricardo's. His work largely consisted in grafting the marginal principle and the principle of substitution on to the Ricardian tradition; and his theory of output and consumption as a whole, as distinct from his theory of the production and distribution of a *given* output, was never separately expounded. Whether he himself felt the need of such a theory, I am not sure. But his immediate successors and followers have certainly dispensed with it and have not, apparently, felt the lack of it. It was in this atmosphere that I was brought up. I taught these doctrines myself and it is only within the last decade that I have been conscious of their insufficiency. In my own thought and development, therefore, this book represents a reaction, a transition away from the English classical (or orthodox) tradition. My emphasis upon this in the following pages and upon the points of my divergence from received doctrine has been regarded in some quarters in England as unduly controversial. But how can one brought up in English economic orthodoxy, indeed a priest of that faith at one time, avoid some controversial emphasis, when he first becomes a Protestant?

Perhaps Japanese readers, however, will neither require nor resist my assaults against the English tradition. We are well aware of the large scale on which English economic writings are read in Japan, but we are not so well informed as to how Japanese opinions regard them. The recent praiseworthy enterprise on the part of the International Economic Circle of Tokyo in reprinting Malthus's 'Principles of Political Economy' as the first volume in the Tokyo Series of Reprints encourages me to think that a book which traces its descent from Malthus rather than Ricardo may be received with sympathy in some quarters at least.

At any rate I am grateful to the *Oriental Economist* for making it possible for me to approach Japanese readers without the extra handicap of a foreign language.

J. M. KEYNES

4 December 1936

Preface to the French Edition

For a hundred years or longer English Political Economy has been dominated by an orthodoxy. That is not to say that an unchanging doctrine has prevailed. On the contrary. There has been a progressive evolution of the doctrine. But its presuppositions, its atmosphere, its method have remained surprisingly the same, and a remarkable continuity has been observable through all the changes. In that orthodoxy, in that continuous transition, I was brought up. I learnt it, I taught it, I wrote it. To those looking from outside I probably still belong to it. Subsequent historians of doctrine will regard this book as in essentially the same tradition. But I myself in writing it, and in other recent work which has led up to it, have felt myself to be breaking away from this orthodoxy, to be in strong reaction against it, to be escaping from something, to be gaining an emancipation. And this state of mind on my part is the explanation of certain faults in the book, in particular its controversial note in some passages, and its air of being addressed too much to the holders of a particular point of view and too little *ad urbem et orbem*. I was wanting to convince my own environment and did not address myself with sufficient directness to outside opinion. Now three years later, having grown accustomed to my new skin and having almost forgotten the smell of my old one, I should, if I were writing afresh, endeavour to free myself from this fault and state my own position in a more clear-cut manner.

I say all this, partly to explain and partly to excuse, myself to French readers. For in France there has been no orthodox tradition with the same authority over contemporary opinion as in my own country. In the United States the position has been much the same as in England. But in France, as in the rest of Europe, there has been no such dominant school since the expiry of the school of French Liberal economists who were in their prime twenty years ago (though they lived to so great an age, long after their influence had passed away, that it fell to my duty, when I first became a youthful editor of the *Economic Journal* to write the obituaries of many of them—Levasseur, Molinari, Leroy-Beaulieu). If Charles Gide had attained to the same influence and authority as Alfred Marshall, your position would have borne more resemblance to ours. As it is, your economists are eclectic, too much (we sometimes think) without deep roots in systematic thought. Perhaps this may make them more easily accessible to what I have to say. But it may also have the result that my readers will sometimes wonder what I am talking about when I speak, with what some of my English critics consider a misuse of language, of the 'classical' school of thought and 'classical' economists. It may, therefore, be helpful to my French readers if I attempt to indicate very briefly what I regard as the main *differentiae* of my approach.

I have called my theory a *general* theory. I mean by this that I am chiefly concerned with the behaviour of the economic system as a whole,—with aggregate incomes, aggregate profits, aggregate output, aggregate employment, aggregate investment, aggregate saving rather than with the incomes, profits, output, employment, investment and saving of particular industries, firms or individuals. And I argue that important mistakes have been made through extending to the system as a whole conclusions which have been correctly arrived at in respect of a part of it taken in isolation.

Let me give examples of what I mean. My contention that for the system as a whole the amount of income which is saved, in the sense that it is not spent on current consumption, is and must necessarily be exactly equal to the amount of net new investment has been considered a paradox and has been the occasion of widespread controversy. The explanation of this is undoubtedly to be found in the fact that this relationship of equality between saving and investment, which necessarily holds good for the system as a whole, does not hold good at all for a particular individual. There is no reason whatever why the new investment for which I am responsible should bear any relation whatever to the amount of my own savings. Quite legitimately we regard an individual's income as independent of what he himself consumes and invests. But this, I have to point out, should not have led us to overlook the fact that the demand arising out of the consumption and investment of one individual is the source of the incomes of other individuals, so that incomes in general are not independent, quite the contrary, of the disposition of individuals to spend and invest; and since in turn the readiness of individuals to spend and invest depends on their incomes, a relationship is set up between aggregate savings and aggregate investment which can be very easily shown, beyond any possibility of reasonable dispute, to be one of exact and necessary equality. Rightly regarded this is a banale conclusion. But it sets in motion a train of thought from which more substantial matters follow. It is shown that, generally speaking, the actual level of output and employment depends, not on the capacity to produce or on the pre-existing level of incomes, but on the current decisions to produce which depend in turn on current decisions to invest and on present expectations of current and prospective consumption. Moreover, as soon as we know the propensity to consume and to save (as I call it), that is to say the result for the community as a whole of the individual psychological inclinations as to how to dispose of given incomes, we can calculate what level of incomes, and therefore what level of output and employment, is in profit-equilibrium with a given level of new investment; out of which develops the doctrine of the Multiplier. Or again, it becomes evident that an increased propensity to save will *ceteris paribus* contract incomes and output; whilst an increased inducement to invest will expand them. We are thus able to analyse the factors which determine the income and output of the system as a whole;—we have, in the most exact sense, a theory of employment. Conclusions emerge from this reasoning which are particularly relevant to the problems of public finance and public policy generally and of the trade cycle.

Another feature, specially characteristic of this book, is the theory of the rate of interest. In recent times it has been held by many economists that the rate of current saving determined the supply of free capital, that the rate of current investment governed the demand for it, and that the rate of interest was, so to speak, the equilibrating price-factor determined by the point of intersection of the supply curve of savings and the demand curve of investment. But if aggregate saving is necessarily and in all circumstances exactly equal to aggregate investment, it is evident that this explanation collapses. We have to search elsewhere for the solution. I find it in the idea that it is the function of the rate of interest to preserve equilibrium, not between the demand and the supply of new capital goods, but between the demand and the supply of money, that is to say between the demand for *liquidity* and the means of satisfying this demand. I am here returning to the doctrine of the older, pre-nineteenth century economists. Montesquieu, for example, saw this truth with considerable clarity, *—Montesquieu who was the real French equivalent of Adam Smith, the greatest of your economists, head and shoulders above the physiocrats in penetration, clear-headedness and good sense (which are the qualities an economist should have). But I must leave it to the text of this book to show how in detail all this works out.

I have called this book the *General Theory of Employment, Interest and Money*; and the third feature to which I may call attention is the treatment of money and prices. The following analysis registers my final escape from the confusions of the Quantity Theory, which once entangled me. I regard the price level as a whole as being determined in precisely the same way as individual prices; that is to say, under the influence of supply and demand. Technical conditions, the level of wages, the extent of unused capacity of plant and labour, and the state of markets and competition determine the supply conditions of individual products and of products as a whole. The decisions of entrepreneurs, which provide the incomes of individual producers and the decisions of those individuals as to the disposition of such incomes determine the conditions. And prices—both individual prices and the price-level— emerge as the resultant of these two factors. Money, and the quantity of money, are not direct influences at this stage of the proceedings. They have

---

* I have particularly in mind Book xxii, chap. 19 of *L'Esprit des lois*.

done their work at an earlier stage of the analysis. The quantity of money determines the supply of liquid resources, and hence the rate of interest, and in conjunction with other factors (particularly that of confidence) the inducement to invest, which in turn fixes the equilibrium level of incomes, output and employment and (at each stage in conjunction with other factors) the price-level as a whole through the influences of supply and demand thus established.

I believe that economics everywhere up to recent times has been dominated, much more than has been understood, by the doctrines associated with the name of J.-B. Say. It is true that his 'law of markets' has been long abandoned by most economists; but they have not extricated themselves from his basic assumptions and particularly from his fallacy that demand is created by supply. Say was implicitly assuming that the economic system was always operating up to its full capacity, so that a new activity was always in substitution for, and never in addition to, some other activity. Nearly all subsequent economic theory has depended on, in the sense that it has required, this same assumption. Yet a theory so based is clearly incompetent to tackle the problems of unemployment and of the trade cycle. Perhaps I can best express to French readers what I claim for this book by saying that in the theory of production it is a final break-away from the doctrines of J.-B. Say and that in the theory of interest it is a return to the doctrines of Montesquieu.

J. M. KEYNES

20 *February* 1939
King's College
Cambridge

Introduction to New Edition

In the spring of 2005 a panel of 'conservative scholars and policy leaders' was asked to identify the most dangerous books of the nineteenth and twentieth centuries. You can get a sense of the panel's leanings by the fact that both Charles Darwin and Betty Friedan ranked high on the list. But *The General Theory of Employment, Interest, and Money* did very well, too. In fact, John Maynard Keynes beat out V. I. Lenin and Frantz Fanon. Keynes, who declared in the book's oft-quoted conclusion that 'soon or late, it is ideas, not vested interests, which are dangerous for good or evil', [384] would probably have been pleased.

Over the past 70 years *The General Theory* has shaped the views even of those who have not heard of it, or who believe they disagree with it. A businessman who warns that falling confidence poses risks for the economy is a Keynesian, whether he knows it or not. A politician who promises that his tax cuts will create jobs by putting spending money in peoples' pockets is a Keynesian, even if he claims to abhor the doctrine. Even self-proclaimed supply-side economists, who claim to have refuted Keynes, fall back on unmistakably Keynesian stories to explain why the economy turned down in a given year.

In this introduction I will address five issues concerning *The General Theory*. First is the book's message—something that ought to be clear from the book itself, but which has often been obscured by those who project their fears or hopes on to Keynes. Second is the question of how Keynes did it: why did he succeed, where others had failed, in convincing the world to accept economic heresy? Third is the question of how much of *The General Theory* remains in today's macroeconomics: are we all Keynesians now, or have we either superseded Keynes's legacy, or, some say, betrayed it? Fourth is the question of what Keynes missed, and why. Finally, I will discuss how Keynes changed economics, and the world.

The message of Keynes

It is probably safe to assume that the 'conservative scholars and policy leaders' who pronounced *The General Theory* one of the most dangerous books of the past two centuries have not read it. But they are sure it is a leftist tract, a call for big government and high taxes. That is what people on the right, and some on the left, too, have said about *The General Theory* from the beginning.

In fact, the arrival of Keynesian economics in American classrooms was delayed by a nasty case of academic McCarthyism. The first introductory textbook to present Keynesian thinking, written by the Canadian economist Lorie Tarshis, was targeted by a right-wing pressure campaign aimed at university trustees. As a result of this campaign, many universities that had planned to adopt the book for their courses cancelled their orders, and sales of the book, which was initially very successful, collapsed. Professors at Yale University, to their credit, continued to assign the book; their reward was to be attacked by the young William F. Buckley for propounding 'evil ideas'.[1]

But Keynes was no socialist—he came to save capitalism, not to bury it. And there's a sense in which *The General Theory* was, given the time it was written, a conservative book. (Keynes himself declared that in some respects his theory was 'moderately conservative in its implications'. [377]) Keynes wrote during a time of mass unemployment, of waste and suffering on an incredible scale. A reasonable man might well have concluded that capitalism had failed, and that only huge institutional changes—perhaps the nationalization of the means of production—could restore economic sanity. Many reasonable people did, in fact, reach that conclusion: large numbers of British and American intellectuals who had no particular antipathy toward markets and private property became socialists during the depression years simply because they saw no other way to remedy capitalism's colossal failures.

Yet Keynes argued that these failures had surprisingly narrow, technical causes. 'We have magneto [alternator] trouble' he wrote in 1930, as the world was plunging into depression.[2] And because Keynes saw the causes of mass unemployment as narrow and technical, he argued that the problem's solution could also be narrow and technical: the system needed a new alternator, but there was no need to replace the whole car. In particular, 'no obvious case is made out for a system of State Socialism which would embrace most of the economic life of the community'. [378] While many of his contemporaries were calling for government takeover of the whole economy, Keynes argued that much less intrusive government policies could ensure adequate effective demand, allowing the market economy to go on as before.

Still, there is a sense in which free-market fundamentalists are right to hate Keynes. If your doctrine says that free markets, left to their own devices, produce the best of all possible worlds, and that government intervention in the economy always makes things worse, Keynes is your enemy. And he is an

---

[1] For a hair-raising account of the coordinated effort to prevent American students from learning Keynesian economics, read David Colander and Harry Landreth's *The Coming of Keynesianism to America*, Edward Elgar, 1996.

[2] 'The Great Slump of 1930', reprinted in *Essays in Persuasion, The Collected Writings of John Maynard Keynes*, Macmillan for the Royal Economic Society, 1972, volume IX, p. 129.

especially dangerous enemy because his ideas have been vindicated so thoroughly by experience.

Stripped down, the conclusions of *The General Theory* might be expressed as four bullet points:

- Economies can and often do suffer from an overall lack of demand, which leads to involuntary unemployment
- The economy's automatic tendency to correct shortfalls in demand, if it exists at all, operates slowly and painfully
- Government policies to increase demand, by contrast, can reduce unemployment quickly
- Sometimes increasing the money supply won't be enough to persuade the private sector to spend more, and government spending must step into the breach

To a modern practitioner of economic policy, none of this—except, possibly, the last point—sounds startling or even especially controversial. But these ideas were not just radical when Keynes proposed them; they were very nearly unthinkable. And the great achievement of *The General Theory* was precisely to make them thinkable.

How Keynes did it

At a guess, most contemporary economists, if they ever actually read *The General Theory*, did so during their student days. Modern academic economics is an endeavor dominated by the new. Often, a whole literature has arisen, flourished, and decayed before the first paper in that literature receives formal publication. Who wants to spend time reading stuff first published 70 years ago?

But *The General Theory* is still worth reading and rereading, not just for what it tells us about the economy, but for what it tells us about the nature of progress in economic thought. Economics students who read Keynes tend to enjoy Keynes's flashes of wit and purple prose, but labour through or skim his elaborate discussions of methodology. But when middle-aged economists—especially those with some experience of the 'struggle of escape' involved in producing a new economic theory—reread Keynes, they see his work from a very different perspective. And they feel a new sense of awe. Parts of the book that once seemed tedious are, one comes to understand, part of a titanic effort to rethink economics, an effort whose success is demonstrated by the fact that so many of Keynes's radical innovations now seem obvious. To really appreciate *The General Theory*, one needs a sense of what Keynes had to go through to get there.

In telling people how to read *The General Theory*, I find it helpful to describe it as a meal that begins with a delectable appetizer and ends with a delightful dessert, but whose main course consists of rather tough meat. It is tempting for readers to dine only on the easily digestible parts of the book, and skip the argument that lies between. But the main course is where the true value of the book lies.

I am not saying that one should skip the fun parts. By all means, read them for the sheer enjoyment, and as a reminder of what Keynes accomplished. In fact, let me say a few words about those parts of the book before I myself get to the hard parts.

Book I is Keynes's manifesto, and for all its academic tone, and even its inclusion of a few equations, it is a thrilling piece of writing. Keynes puts you, the professional economist—for *The General Theory* was, above all, a book written for knowledgeable insiders—on notice that he is going to refute everything you thought you knew about employment. In just a few pages he convincingly shows that the then conventional view about the relationship between wages and employment involves a basic fallacy of composition: 'In assuming that the wage bargain determines the real wage the classical school have slipt into an illicit assumption.' [13]. From this, he quickly shows that the conventional view that wage cuts were the route to full employment made no sense given the realities of the time. And in just a few more pages he lays out enough of his own theory to suggest the breathtaking conclusion that the Great Depression then afflicting the world was not only solvable, but easily solvable.

It is a bravura performance. Modern readers who stop after Book I, however, without slogging through the far denser chapters that follow, get a sense of Keynes's audacity, but not of how he earned the right to that audacity.

Book VI, at the opposite end of *The General Theory*, really is a kind of dessert course. Keynes, the hard work of creating macroeconomics as we know it behind him, kicks up his heels and has a little fun. In particular, the final two chapters of *The General Theory*, though full of interesting ideas, have an impish quality. Keynes tells us that the famous victory of free trade over protectionism may have been won on false pretenses—that the mercantilists had a point. He tells us that the 'euthanasia of the rentier' [376] may be imminent, because thrift no longer serves a social function. Did he really believe these things, or was he simply enjoying tweaking the noses of his colleagues? Probably some of both.

Again, Book VI is a great read, although it has not stood the test of time nearly as well as Book I. But the same caution applies: by all means, read Keynes's speculations on the virtues of mercantilism and the vanishing need for thrift, but remember that the tough stuff in Books II through V is what gave him the right to speculate.

So now let us talk about the core of the book, and what it took for Keynes to write it.

There is no shortage of people willing to challenge conventional economic wisdom; I must receive at least one new book every month that claims to do just that. The vast majority of these books' authors, however, do not understand enough about existing economic theory to mount a credible challenge.

Keynes, by contrast, was deeply versed in the economic theory of his time, and understood the power of that body of theory. 'I myself', he wrote in the preface, 'held with conviction for many years the very theories which I now attack, and am not, I think, unaware of their strong points'. [p. xxi] He knew that he had to offer a coherent, carefully reasoned challenge to the reigning orthodoxy to change peoples' minds. In Book I, as Keynes gives us a first taste of what he is going to do, he writes of Malthus, whose intuition told him that general failures of demand were possible, but had no model to back that intuition: '[S]ince Malthus was unable to explain clearly (apart from an appeal to the facts of common observation) how and why effective demand could be deficient or excessive, he failed to provide an alternative construction; and Ricardo conquered England as completely as the Holy Inquisition conquered Spain'. [32]

That need to 'provide an alternative construction' explains many of the passages in *The General Theory* that, 70 years later, can seem plodding or even turgid. In particular, it explains Book II, which most modern readers probably skip. Why devote a whole chapter to 'the choice of units', which does not seem to have much to do with Keynes's grand vision? Why devote two more chapters to defining the meaning of income, savings, and investment? For the same reason that economists working on applications of increasing returns to international trade and economic growth theory in the 1980s lavished many pages on the details of product differentiation and monopolistic competition. These details had nothing much to do with the fundamental ideas behind the new theories. But the details were crucial to producing the buttoned-down models economists needed to clarify their thoughts and explain those thoughts to others. When you are challenging a long-established orthodoxy, the vision thing does not work unless you are very precise about the details.

Keynes's appreciation of the power of the reigning orthodoxy also explains the measured pace of his writing. 'The composition of this book', wrote Keynes in the preface, 'has been for the author a long struggle of escape, and so must the reading of it be....' [p. xxiii] Step by step, Keynes set out to liberate economists from the intellectual confines that left them unable to deal with the Great Depression, confines created for the most part by what Keynes dubbed 'classical economics'.

Keynes's struggle with classical economics was much more difficult than we can easily imagine today. Modern macroeconomics textbooks usually contain a discussion of something they call the 'classical model' of the price level. But that model offers far too flattering a picture of the classical economics Keynes had to escape from. What we call the classical model today is really a post-Keynesian attempt to rationalize pre-Keynesian views. Change one assumption in our so-called classical model, that of perfect wage flexibility, and it turns back into *The General Theory*. If that had been all Keynes had to contend with, *The General Theory* would have been an easy book to write.

The real classical model, as Keynes described it, was something much harder to fix. It was, essentially, a model of a barter economy, in which money and nominal prices do not matter, with a monetary theory of the price level appended in a non-essential way, like a veneer on a tabletop. It was a model in which Say's Law applied: supply automatically creates its own demand, because income must be spent. And it was a model in which the interest rate was purely a matter of the supply and demand for funds, with no possible role for money or monetary policy. It was, as I said, a model in which ideas we now take for granted were literally unthinkable.

If the classical economics Keynes confronted had been what we call the classical model nowadays, he would not have had to write Book V of *The General Theory*, 'Money-wages and prices'. In that book Keynes confronts naïve beliefs about how a fall in wages can increase employment, beliefs that were prevalent among economists when he wrote, but play no role in the model we now call 'classical'.

So the crucial innovation in *The General Theory* is not, as a modern macroeconomist tends to think, the idea that nominal wages are sticky. It is the demolition of Say's Law and the classical theory of the interest rate in Book IV, 'The inducement to invest'. One measure of how hard it was for Keynes to divest himself of Say's Law is that to this day some people deny what Keynes realized—that the 'law' is, at best, a useless tautology when individuals have the option of accumulating money rather than purchasing real goods and services. Another measure of Keynes's achievement may be hard to appreciate unless you've taught introductory macroeconomics: how do you explain to students how the central bank can reduce the interest rate by increasing the money supply, even though the interest rate is the price at which the supply of loans is equal to the demand? It is not easy to explain even when you know the answer; think how much harder it was for Keynes to arrive at the right answer in the first place.

But the classical model wasn't the only thing Keynes had to escape from. He also had to break free of the business cycle theory of the day.

There was not, of course, anything like a fully-worked out model of recessions and recoveries. But it is instructive to compare *The General Theory* with Gottfried Haberler's *Prosperity and Depression*, written at roughly the same time, which was a League of Nations-sponsored attempt to systematize and synthesize what the economists of the time had to say about the subject.[4] What is striking about Haberler's book, from a modern perspective—aside from the absence of any models—is that he was trying to answer the wrong question. Like most macroeconomic theorists before Keynes, Haberler believed that the crucial thing was to explain the economy's dynamics, to explain why booms are followed by busts, rather than to explain how mass unemployment is possible in the first place. And Haberler's book, like much business cycle writing at the time, seems more preoccupied with the excesses of the boom than with the mechanics of the bust. Although Keynes speculated about the causes of the business cycle in Chapter 22 of *The General Theory*, those speculations were peripheral to his argument. Instead, Keynes saw it as his job to explain why the economy sometimes operates far below full employment. That is, *The General Theory* for the most part offers a static model, not a dynamic model—a picture of an economy stuck in depression, not a story about how it got there. So Keynes actually chose to answer a more limited question than most people writing about business cycles at the time.

Indeed, most of Book II of *The General Theory* is a manifesto on behalf of Keynes's strategic decision to limit the question. Where pre-Keynesian business cycle theory told complex, confusing stories about disequilibrium, Chapter 5 makes the case for thinking of an underemployed economy as being in a sort of equilibrium in which short-term expectations about sales are, in fact, fulfilled. Chapter 6 and Chapter 7 argue for replacing all the talk of forced savings, excess savings, and so on that was prevalent in pre-Keynesian business cycle theory—talk that stressed, in a confused way, the idea of disequilibrium in the economy—with the simple accounting identity that savings equal investment.

And Keynes's limitation of the question was powerfully liberating. Rather than getting bogged down in an attempt to explain the dynamics of the business cycle—a subject that remains contentious to this day—Keynes focused on a question that could be answered. And that was also the question that most needed an answer: given that overall demand is depressed—never mind why—how can we create more employment?

A side benefit of this simplification was that it freed Keynes and the rest of us from the seductive but surely false notion of the business cycle as morality play, of an economic slump as a necessary purgative after the excesses of a boom. By analysing how the economy stays depressed, rather than trying to explain how it became depressed in the first place, Keynes helped bury the notion that there is something redemptive about economic suffering.

---

[4] Gottfried Haberler, *Prosperity and Depression*, League of Nations, 1937.

The General Theory, then, is a work of informed, disciplined, intellectual radicalism. It transformed the way everyone, including Keynes's intellectual opponents, thought about the economy. But that raises a contentious question: are we, in fact, all Keynesians now?

Mr. Keynes and the moderns

There is a widespread impression among modern macroeconomists that we have left Keynes behind, for better or for worse. But that impression, I'd argue, is based either on a misreading or a non-reading of *The General Theory*. Let's start with the non-readers, a group that included me during the several decades that passed between my first and second readings of *The General Theory*.

If you do not read Keynes himself, but only read his work as refracted through various interpreters, it is easy to imagine that *The General Theory* is much cruder than it is. Even professional economists, who know that Keynes was not a raving socialist, tend to think that *The General Theory* is largely a manifesto proclaiming the need for deficit spending, and that it belittles monetary policy. If that were really true, *The General Theory* would be a very dated book. These days economic stabilization is mainly left up to technocrats in central banks, who move interest rates up and down through their control of the money supply; the use of public works spending to prop up employment is generally considered unnecessary. To put it crudely, if you imagine that Keynes was dismissive of monetary policy, it is easy to imagine that Milton Friedman in some sense refuted or superseded Keynes by showing that money matters.

The impression that *The General Theory* failed to give monetary policy its due may have been reinforced by John Hicks, whose 1937 review essay 'Mr. Keynes and the classics' is probably more read by economists these days than *The General Theory* itself. In that essay Hicks interpreted *The General Theory* in terms of two curves, the IS curve, which can be shifted by changes in taxes and spending, and the LM curve, which can be shifted by changes in the money supply. And Hicks seemed to imply that Keynesian economics applies only when the LM curve is flat, so that changes in the money supply do not affect interest rates, while classical macroeconomics applies when the LM curve is upward-sloping.

But in this implication Hicks was both excessively kind to the classics and unfair to Keynes. I have already pointed out that the macroeconomic doctrine from which Keynes had to escape was much cruder and more confused than the doctrine we now call the 'classical model'. Let me add that *The General Theory* does not dismiss or ignore monetary policy. Keynes discusses at some length how changes in the quantity of money can affect the rate of interest, and through the rate of interest affect aggregate demand. In fact, the modern theory of how monetary policy works is essentially that laid out in *The General Theory*.

Yet it is fair to say that *The General Theory* is pervaded by skepticism about whether merely adding to the money supply is enough to restore full employment. This was not because Keynes was ignorant of the potential role of monetary policy. Rather, it was an empirical judgment on his part: *The General Theory* was written in an economy with interest rates already so low that there was little an increase in the money supply could do to push them lower.

Many of today's most prominent macroeconomists came of intellectual age during the 1970s and 1980s, when interest rates were consistently above 5 percent and sometimes in double digits. Under those conditions there was no reason to doubt the effectiveness of monetary policy, no reason to worry that the central bank could fail in efforts to drive down interest rates and thereby increase demand. But *The General Theory* was written in a very different monetary environment, one in which interest rates stayed close to zero for an extended period.

Modern macroeconomists do not have to theorize about what happens to monetary policy in such an environment, or even plumb the depths of economic history, because we have a striking recent example to contemplate. There are hopes as I write this that the Japanese economy may finally be staging a sustained recovery, but from the early 1990s at least through 2004 Japan was in much the same monetary state that the U.S. and U.K. economies were in during the 1930s. Short-term interest rates were close to zero, long-term rates were at historical lows, yet private investment spending remained insufficient to bring the economy out of deflation. In that environment, monetary policy was just as ineffective as Keynes described. Attempts by the Bank of Japan to increase the money supply simply added to already ample bank reserves and public holdings of cash while doing nothing to stimulate the economy. (A Japanese joke from the late 90s said that safes were the only product consumers were buying.) And when the Bank of Japan found itself impotent, the government of Japan turned to large public works projects to prop up demand.

Keynes made it clear that his skepticism about the effectiveness of monetary policy was a contingent proposition, not a statement of a general principle. In the past, he believed, things had been otherwise. 'There is evidence that for a period of almost one hundred and fifty years the long-run typical rate of interest in the leading financial centres was about 5 percent, and the gilt-edged rate between 3 and $3^{1}/_{2}$ percent; and that these rates were modest enough to encourage a rate of investment consistent with an average of employment which was not intolerably low'. [307–308] In that environment, he believed, 'a tolerable level of unemployment could be attained on the average of one or two or three decades merely by assuring an adequate supply of money in terms of wage-units'. [309] In other words, monetary policy had worked in the past—but not now.

Now it is true that Keynes believed, wrongly, that the conditions of the 1930s would persist indefinitely—indeed, that the marginal efficiency of capital was falling to the point that the euthanasia of rentiers was in view. I will mention why he was wrong in a moment.

Before I get there, however, let me consider talk about an alternative view. This view agrees with those who say that modern macroeconomics owes little to Keynes. But rather than arguing that we have superseded Keynes, this view says that we have misunderstood him. That is, some economists insist that we've lost the true Keynesian path—that modern macroeconomic theory, which reduces Keynes to a static equilibrium model, and tries to base as much of that model as possible on rational choice, is a betrayal of Keynesian thinking.

Is this right? On the issue of rational choice, it's true that compared with any modern exposition of macroeconomics, *The General Theory* contains very little discussion of maximization and a lot of behavioural hypothesizing. Keynes's emphasis on the non-rational roots of economic behaviour is most quotable when he writes of financial market speculation, 'where we devote our intelligences to anticipating what average opinion expects average opinion to be'. [156] But it is most notable, from a modern perspective, in his discussion of the consumption function. Attempts to model consumption behaviour in terms of rational choice were one of the main themes of macroeconomics after Keynes. But Keynes's consumption function, as laid out in Book III, is grounded in psychological observation rather than intertemporal optimization.

This raises two questions. First, was Keynes right to eschew maximizing theory? Second, did his successors betray his legacy by bringing maximization back in?

The answer to the first question is, it depends. Keynes was surely right that there is a strong non-rational element in economic behavior. The rise of behavioral economics and behavioral finance is a belated recognition by the profession of this fact. On the other hand, some of Keynes's attempted generalizations about behavior now seem excessively facile and misleading in important ways. In particular, he argued on psychological grounds that the average savings rate would rise with per capita income (see p. 97.) That has turned out to be not at all the case.

But the answer to the second question, I would argue, is clearly no. Yes, Keynes was a shrewd observer of economic irrationality, a behavioural economist before his time, who had a lot to say about economic dynamics. Yes, *The General Theory* is full of witty passages about investing as a game of musical chairs, about animal spirits, and so on. But *The General Theory* is not primarily a book about the unpredictability and irrationality of economic actors. Keynes emphasizes the relative stability of the relationship between income and consumer spending; trying to ground that stability in rational choice may be wrong-headed, but it does not undermine his intent. And while Keynes didn't think much of the rationality of business behaviour, one of the key strategic decisions he made, as I have already suggested, was to push the whole question of why investment rises and falls into the background.

What about equilibrium? Let me offer some fighting words: to interpret Keynes in terms of static equilibrium models is no betrayal, because what Keynes mainly produced was indeed a static equilibrium model. The essential story laid out in *The General Theory* is that liquidity preference determines the rate of interest; given the rate of interest, the marginal efficiency of capital determines the rate of investment; and employment is determined by the point at which the value of output is equal to the sum of investment and consumer spending. '[G]iven the propensity to consume and the rate of new investment, there will be only one level of employment consistent with equilibrium'. [28]

Let me address one issue in particular: did Paul Samuelson, whose 1948 text-book introduced the famous 45-degree diagram to explain the multiplier, mis-represent what Keynes was all about? There are commentators who insist passionately that Samuelson defiled the master's thought. Yet it is hard to see any significant difference between Samuelson's formulation and Keynes's own equation for equilibrium employment, right there in Chapter 3: $\phi(N) - \chi(N) = D_2$ [29]. Represented graphically, Keynes's version looks a lot like Samuelson's diagram; quantities are measured in wage units rather than constant dollars, and the nifty 45-degree feature is absent, but the logic is exactly the same.

The bottom line, then, is that we really are all Keynesians now. A very large part of what modern macroeconomists do derives directly from *The General Theory*; the framework Keynes introduced holds up very well to this day.

Yet there were, of course, important things that Keynes missed or failed to anticipate.

What Keynes missed

The strongest criticism one can make of *The General Theory* is that Keynes mistook an episode for a trend. He wrote in a decade when even a near-zero interest rate wasn't low enough to restore full employment, and brilliantly explained the implications of that fact—in particular, the trap in which the Bank of England and the Federal Reserve found themselves, unable to create employment no matter how much they tried to increase the money supply.

He knew that matters had not always been thus. But he believed, wrongly, that the monetary environment of the 1930s would be the norm from then on.

Japan aside, the monetary conditions of the 1930s have not made a re-appearance. In the United States the era of ultra-low interest rates ended in the 1950s, and has never returned (although the U.S. had a near-Japan experience in 2002–2003.) Yet the United States has, in general, succeeded in achieving adequate levels of effective demand. The British experience has been similar. And although there is large-scale unemployment in continental Europe, that unemployment seems to have more to do with supply-side issues than with sheer lack of demand.

Why was Keynes wrong?

Part of the answer is that he underestimated the ability of mature economies to stave off diminishing returns. Keynes's 'euthanasia of the rentier' was predicated on the presumption that as capital accumulates, profitable private investment projects become harder to find, so that the marginal efficiency of capital declines. In interwar Britain, with the heroic era of industrialization behind it, that view may have seemed reasonable. But after World War II a combination of technological progress and revived population growth opened up many new investment opportunities. And even though there have lately been warnings of a 'global savings glut', the euthanasia of the rentier does not seem imminent.

But there is an even more important factor that has kept interest rates relatively high, and monetary policy effective: persistent inflation, which has become embedded in expectations, and is reflected in higher interest rates than we would have if the public expected stable prices. Inflation was, of course, much higher in the 1970s and even the 1980s than it is today. Yet expectations of inflation still play a powerful role in keeping interest rates safely away from zero. For example, at the time of writing the interest rate on 20-year U.S. government bonds was 4.7%; the interest rate on 20-year 'indexed' bonds, whose return is protected from inflation, was only 2.1%. This tells us that even now, when inflation is considered low, most of the 20-year rate reflects expected inflation rather than expected real returns.

The irony is that persistent inflation, which makes *The General Theory* seem on the surface somewhat less directly relevant to our time than it would in the absence of that inflation, can be attributed in part to Keynes's influence, for better or worse. For worse: the inflationary take-off of the 1970s was partly caused by expansionary monetary and fiscal policy, adopted by Keynes-influenced governments with unrealistic employment goals. (I am thinking in particular of Edward Heath's 'dash for growth' in the UK and the Burns–Nixon boom in the US.) For better: both the Bank of England, explicitly, and the Federal Reserve, implicitly, have a deliberate strategy of encouraging persistent low but positive inflation, precisely to avoid finding themselves in the trap Keynes diagnosed.

Keynes did not foresee a future of persistent inflation (nor did anyone else at the time.) This meant that he was excessively pessimistic about the future prospects for monetary policy. It also meant that he never addressed the policy problems posed by persistent inflation, which preoccupied macro-economists in the 70s and 80s, and led some to proclaim a crisis in economic theory. (In fact, some of the models widely used today to explain the persistence of inflation even in the face of unemployment, notably 'overlapping contracts' models that stress the uncoordinated nature of wage settlements, are quite consistent in spirit with what Keynes had to say about wage determination.) But failure to address problems nobody imagined in the 1930s can hardly be considered a flaw in Keynes's analysis. And now that inflation has subsided, Keynes looks highly relevant again.

The economist as saviour

As an intellectual achievement, *The General Theory* ranks with only a handful of other works in economics—the tiny set of books that transformed our perception of the world, so that once people became aware of what those books had to say they saw everything differently. Adam Smith did that in *The Wealth of Nations*: suddenly the economy was not just a collection of people getting and spending, it was a self-regulating system in which each individual 'is led by an invisible hand to promote an end which was no part of his intention'. *The General Theory* is in the same league: suddenly the idea that mass unemployment is the result of inadequate demand, long a fringe heresy, became completely comprehensible, indeed obvious.

What makes *The General Theory* truly unique, however, is that it combined towering intellectual achievement with immediate practical relevance to a global economic crisis. The second volume of Robert Skidelsky's biography of Keynes is titled *The Economist as Saviour*, and there's not a bit of hyperbole involved. Until *The General Theory*, sensible people regarded mass unemployment as a problem with complex causes, and no easy solution other than the replacement of markets with government control. Keynes showed that the opposite was true: mass unemployment had a simple cause, inadequate demand, and an easy solution, expansionary fiscal policy.

It would be a wonderful story if *The General Theory* showed the world the way out of depression. Alas for romance, that is not quite what happened. The giant public works programme that restored full employment, otherwise known as the Second World War, was launched for reasons unrelated to macroeconomic theory. But Keynesian theory explained why war spending did what it did, and helped governments ensure that the postwar world didn't slip back into depression. And who is to say that depression-like conditions would not have returned if governments, taking their guidance from Keynesian economics, had not responded to recessions with expansionary policies?

There has been nothing like Keynes's achievement in the annals of social science. Perhaps there cannot be. Keynes was right about the problem of his day: the world economy had magneto trouble, and all it took to get the economy going again was a surprisingly narrow, technical fix. But most economic problems probably do have complex causes and do not have easy solutions. Maybe there are narrow, technical solutions to the economic problems of today's world, from lagging development in Latin America to soaring inequality in the United States, and we are just waiting for the next Keynes to discover them. But at the moment it does not seem likely.

One thing is certain: if there is another Keynes out there, he or she will be someone who shares Keynes's most important qualities. Keynes was a consummate intellectual insider, who understood the prevailing economic ideas of his day as well as anyone. Without that base of knowledge, and the skill in argumentation that went with it, he would not have been able to mount such a devastating critique of economic orthodoxy. Yet he was at the same time a daring radical, willing to consider the possibility that some of the fundamental assumptions of the economics he had been taught were wrong.

Those qualities allowed Keynes to lead economists, and the world, into the light—for *The General Theory* is nothing less than an epic journey out of intellectual darkness. That, as much as its continuing relevance to economic policy, is what makes it a book for the ages. Read it, and marvel.

Paul Krugman

BOOK I
INTRODUCTION

Chapter 1

THE GENERAL THEORY

I have called this book the *General Theory of Employment, Interest and Money*, placing the emphasis on the prefix *general*. The object of such a title is to contrast the character of my arguments and conclusions with those of the *classical*[1] theory of the subject, upon which I was brought up and which dominates the economic thought, both practical and theoretical, of the governing and academic classes of this generation, as it has for a hundred years past. I shall argue that the postulates of the classical theory are applicable to a special case only and not to the general case, the situation which it assumes being a limiting point of the possible positions of equilibrium. Moreover, the characteristics of the special case assumed by the classical theory happen not to be those of the economic society in which we actually live, with the result that its teaching is misleading and disastrous if we attempt to apply it to the facts of experience.

[1] 'The classical economists' was a name invented by Marx to cover Ricardo and James Mill and their *predecessors*, that is to say for the founders of the theory which culminated in the Ricardian economics. I have become accustomed, perhaps perpetrating a solecism, to include in 'the classical school' the *followers* of Ricardo, those, that is to say, who adopted and perfected the theory of the Ricardian economics, including (for example) J. S. Mill, Marshall, Edgeworth and Prof. Pigou.

Chapter 2

THE POSTULATES OF THE
CLASSICAL ECONOMICS

Most treatises on the theory of value and production are primarily concerned with the distribution of a *given* volume of employed resources between different uses and with the conditions which, assuming the employment of this quantity of resources, determine their relative rewards and the relative values of their products.[1]

The question, also, of the volume of the *available* resources, in the sense of the size of the employable population, the extent of natural wealth and the accumulated capital equipment, has often been treated descriptively. But the pure theory of what determines the *actual employment* of the available resources has seldom been examined in great detail. To say that it has not been examined at all would, of course, be absurd. For every discussion concerning fluctuations of employment, of which there have been many, has been concerned with it. I mean, not that the topic has been overlooked, but that the fundamental theory

[1] This is in the Ricardian tradition. For Ricardo expressly repudiated any interest in the *amount* of the national dividend, as distinct from its distribution. In this he was assessing correctly the character of his own theory. But his successors, less clear-sighted, have used the classical theory in discussions concerning the causes of wealth. *Vide* Ricardo's letter to Malthus of October 9, 1820: 'Political Economy you think is an enquiry into the nature and causes of wealth—I think it should be called an enquiry into the laws which determine the division of the produce of industry amongst the classes who concur in its formation. No law can be laid down respecting quantity, but a tolerably correct one can be laid down respecting proportions. Every day I am more satisfied that the former enquiry is vain and delusive, and the latter only the true objects of the science.'

4

underlying it has been deemed so simple and obvious that it has received, at the most, a bare mention.[1]

I

The classical theory of employment—supposedly simple and obvious—has been based, I think, on two fundamental postulates, though practically without discussion, namely:

I. *The wage is equal to the marginal product of labour*

That is to say, the wage of an employed person is equal to the value which would be lost if employment were to be reduced by one unit (after deducting any other costs which this reduction of output would avoid); subject, however, to the qualification that the equality may be disturbed, in accordance with certain principles, if competition and markets are imperfect.

II. *The utility of the wage when a given volume of labour is employed is equal to the marginal disutility of that amount of employment.*

That is to say, the real wage of an employed person is that which is just sufficient (in the estimation of the employed persons themselves) to induce the volume of labour actually employed to be forthcoming; subject to the qualification that the equality for each individual unit of labour may be disturbed by combination between employable units analogous to the imperfections

---

[1] For example, Prof. Pigou in the *Economics of Welfare* (4th ed. p. 127) writes (my italics): 'Throughout this discussion, except when the contrary is expressly stated, the fact that some resources are generally unemployed against the will of the owners is ignored. *This does not affect the substance of the argument*, while it simplifies its exposition.' Thus, whilst Ricardo expressly disclaimed any attempt to deal with the amount of the national dividend as a whole, Prof. Pigou, in a book which is specifically directed to the problem of the national dividend, maintains that the same theory holds good when there is some involuntary unemployment as in the case of full employment.

5

of competition which qualify the first postulate. Disutility must be here understood to cover every kind of reason which might lead a man, or a body of men, to withhold their labour rather than accept a wage which had to them a utility below a certain minimum.

This postulate is compatible with what may be called 'frictional' unemployment. For a realistic interpretation of it legitimately allows for various inexactnesses of adjustment which stand in the way of continuous full employment: for example, unemployment due to a temporary want of balance between the relative quantities of specialised resources as a result of miscalculation or intermittent demand; or to time-lags consequent on unforeseen changes; or to the fact that the change-over from one employment to another cannot be effected without a certain delay, so that there will always exist in a non-static society a proportion of resources unemployed 'between jobs'. In addition to 'frictional' unemployment, the postulate is also compatible with 'voluntary' unemployment due to the refusal or inability of a unit of labour, as a result of legislation or social practices or of combination for collective bargaining or of slow response to change or of mere human obstinacy, to accept a reward corresponding to the value of the product attributable to its marginal productivity. But these two categories of 'frictional' unemployment and 'voluntary' unemployment are comprehensive. The classical postulates do not admit of the possibility of the third category, which I shall define below as 'involuntary' unemployment.

Subject to these qualifications, the volume of employed resources is duly determined, according to the classical theory, by the two postulates. The first gives us the demand schedule for employment, the second gives us the supply schedule; and the amount of employment is fixed at the point where the utility of the marginal product balances the disutility of the marginal employment.

It would follow from this that there are only four possible means of increasing employment:

(*a*) An improvement in organisation or in foresight which diminishes 'frictional' unemployment;

(*b*) a decrease in the marginal disutility of labour, as expressed by the real wage for which additional labour is available, so as to diminish 'voluntary' unemployment;

(*c*) an increase in the marginal physical productivity of labour in the wage-goods industries (to use Professor Pigou's convenient term for goods upon the price of which the utility of the money-wage depends);

or (*d*) an increase in the price of non-wage-goods compared with the price of wage-goods, associated with a shift in the expenditure of non-wage-earners from wage-goods to non-wage-goods.

This, to the best of my understanding, is the substance of Professor Pigou's *Theory of Unemployment* —the only detailed account of the classical theory of employment which exists.[1]

II

Is it true that the above categories are comprehensive in view of the fact that the population generally is seldom doing as much work as it would like to do on the basis of the current wage? For, admittedly, more labour would, as a rule, be forthcoming at the existing money-wage if it were demanded.[2] The classical school reconcile this phenomenon with their second postulate by arguing that, while the demand for labour

[1] Prof. Pigou's *Theory of Unemployment* is examined in more detail in the Appendix to Chapter 19 below.
[2] Cf. the quotation from Prof. Pigou above, p. 5, footnote.

at the existing money-wage may be satisfied before everyone willing to work at this wage is employed, this situation is due to an open or tacit agreement amongst workers not to work for less, and that if labour as a whole would agree to a reduction of money-wages more employment would be forthcoming. If this is the case, such unemployment, though apparently involuntary, is not strictly so, and ought to be included under the above category of 'voluntary' unemployment due to the effects of collective bargaining, etc.

This calls for two observations, the first of which relates to the actual attitude of workers towards real wages and money-wages respectively and is not theoretically fundamental, but the second of which is fundamental.

Let us assume, for the moment, that labour is not prepared to work for a lower money-wage and that a reduction in the existing level of money-wages would lead, through strikes or otherwise, to a withdrawal from the labour market of labour which is now employed. Does it follow from this that the existing level of real wages accurately measures the marginal disutility of labour? Not necessarily. For, although a reduction in the existing money-wage would lead to a withdrawal of labour, it does not follow that a fall in the value of the existing money-wage in terms of wage-goods would do so, if it were due to a rise in the price of the latter. In other words, it may be the case that within a certain range the demand of labour is for a minimum money-wage and not for a minimum real wage. The classical school have tacitly assumed that this would involve no significant change in their theory. But this is not so. For if the supply of labour is not a function of real wages as its sole variable, their argument breaks down entirely and leaves the question of what the actual employment will be quite indeterminate.[1] They do not seem to have realised that, unless the supply of labour is a function of real wages alone,

[1] This point is dealt with in detail in the Appendix to chapter 19 below.

their supply curve for labour will shift bodily with every movement of prices. Thus their method is tied up with their very special assumptions, and cannot be adapted to deal with the more general case.

Now ordinary experience tells us, beyond doubt, that a situation where labour stipulates (within limits) for a money-wage rather than a real wage, so far from being a mere possibility, is the normal case. Whilst workers will usually resist a reduction of money-wages, it is not their practice to withdraw their labour whenever there is a rise in the price of wage-goods. It is sometimes said that it would be illogical for labour to resist a reduction of money-wages but not to resist a reduction of real wages. For reasons given below (p. 14), this might not be so illogical as it appears at first; and, as we shall see later, fortunately so. But, whether logical or illogical, experience shows that this is how labour in fact behaves.

Moreover, the contention that the unemployment which characterises a depression is due to a refusal by labour to accept a reduction of money-wages is not clearly supported by the facts. It is not very plausible to assert that unemployment in the United States in 1932 was due either to labour obstinately refusing to accept a reduction of money-wages or to its obstinately demanding a real wage beyond what the productivity of the economic machine was capable of furnishing. Wide variations are experienced in the volume of employment without any apparent change either in the minimum real demands of labour or in its productivity. Labour is not more truculent in the depression than in the boom—far from it. Nor is its physical productivity less. These facts from experience are a *prima facie* ground for questioning the adequacy of the classical analysis.

It[1] would be interesting to see the results of a statistical enquiry into the actual relationship between

---

[1] See Appendix 3 below for further discussion on this point. [Ed.]

changes in money-wages and changes in real wages. In the case of a change peculiar to a particular industry one would expect the change in real wages to be in the same direction as the change in money-wages. But in the case of changes in the general level of wages, it will be found, I think, that the change in real wages associated with a change in money-wages, so far from being usually in the same direction, is almost always in the opposite direction. When money-wages are rising, that is to say, it will be found that real wages are falling; and when money-wages are falling, real wages are rising. This is because, in the short period, falling money-wages and rising real wages are each, for independent reasons, likely to accompany decreasing employment; labour being readier to accept wage-cuts when employment is falling off, yet real wages inevitably rising in the same circumstances on account of the increasing marginal return to a given capital equipment when output is diminished.

If, indeed, it were true that the existing real wage is a minimum below which more labour than is now employed will not be forthcoming in any circumstances, involuntary unemployment, apart from frictional unemployment, would be non-existent. But to suppose that this is invariably the case would be absurd. For more labour than is at present employed is usually available at the existing money-wage, even though the price of wage-goods is rising and, consequently, the real wage falling. If this is true, the wage-goods equivalent of the existing money-wage is not an accurate indication of the marginal disutility of labour, and the second postulate does not hold good.

But there is a more fundamental objection. The second postulate flows from the idea that the real wages of labour depend on the wage bargains which labour makes with the entrepreneurs. It is admitted, of course, that the bargains are actually made in terms of money, and even that the real wages acceptable to labour are not altogether independent of what the corresponding money-wage happens to be. Nevertheless it is the money-wage thus arrived at which is held to determine the real wage. Thus the classical theory assumes that it is always open to labour to reduce its real wage by accepting a reduction in its money-wage. The postulate that there is a tendency for the real wage to come to equality with the marginal disutility of labour clearly presumes that labour itself is in a position to decide the real wage for which it works, though not the quantity of employment forthcoming at this wage.

The traditional theory maintains, in short, *that the wage bargains between the entrepreneurs and the workers determine the real wage*; so that, assuming free competition amongst employers and no restrictive combination amongst workers, the latter can, if they wish, bring their real wages into conformity with the marginal disutility of the amount of employment offered by the employers at that wage. If this is not true, then there is no longer any reason to expect a tendency towards equality between the real wage and the marginal disutility of labour.

The classical conclusions are intended, it must be remembered, to apply to the whole body of labour and do not mean merely that a single individual can get employment by accepting a cut in money-wages which his fellows refuse. They are supposed to be equally applicable to a closed system as to an open system, and are not dependent on the characteristics of an open system or on the effects of a reduction of money-wages in a single country on its foreign trade, which lie, of course, entirely outside the field of this discussion. Nor are they based on indirect effects due to a lower wages-bill in terms of money having certain reactions on the banking system and the state of credit, effects which we shall examine in detail in chapter 19. They are based on the belief that in a closed system a reduction in the general level of money-wages will be accompanied, at any rate in the short period and subject only to minor qualifications, by some, though not always a proportionate, reduction in real wages.

Now the assumption that the general level of real wages depends on the money-wage bargains between the employers and the workers is not obviously true. Indeed it is strange that so little attempt should have been made to prove or to refute it. For it is far from being consistent with the general tenor of the classical theory, which has taught us to believe that prices are governed by marginal prime cost in terms of money and that money-wages largely govern marginal prime cost. Thus if money-wages change, one would have expected the classical school to argue that prices would change in almost the same proportion, leaving the real wage and the level of unemployment practically the same as before, any small gain or loss to labour being at the expense or profit of other elements of marginal cost which have been left unaltered.[1] They seem, however, to have been diverted from this line of thought, partly by the settled conviction that labour is in a position to determine its own real wage and partly, perhaps, by preoccupation with the idea that prices depend on the quantity of money. And the belief in the proposition that labour is always in a position to determine its own real wage, once adopted, has been maintained by its being confused with the proposition that labour is always in a position to determine what real wage shall correspond to *full* employment, i.e. the *maximum* quantity of employment which is compatible with a given real wage.

To sum up: there are two objections to the second postulate of the classical theory. The first relates to the actual behaviour of labour. A fall in real wages due

[1] This argument would, indeed, contain, to my thinking, a large element of truth, though the complete results of a change in money-wages are more complex, as we shall show in chapter 19 below.

to a rise in prices, with money-wages unaltered, does not, as a rule, cause the supply of available labour on offer at the current wage to fall below the amount actually employed prior to the rise of prices. To suppose that it does is to suppose that all those who are now unemployed though willing to work at the current wage will withdraw the offer of their labour in the event of even a small rise in the cost of living. Yet this strange supposition apparently underlies Professor Pigou's *Theory of Unemployment*,[1] and it is what all members of the orthodox school are tacitly assuming.

But the other, more fundamental, objection, which we shall develop in the ensuing chapters, flows from our disputing the assumption that the general level of real wages is directly determined by the character of the wage bargain. In assuming that the wage bargain determines the real wage the classical school have slipt in an illicit assumption. For there may be *no* method available to labour as a whole whereby it can bring the wage-goods equivalent of the general level of money-wages into conformity with the marginal disutility of the current volume of employment. There may exist no expedient by which labour as a whole can reduce its *real* wage to a given figure by making revised *money* bargains with the entrepreneurs. This will be our contention. We shall endeavour to show that primarily it is certain other forces which determine the general level of real wages. The attempt to elucidate this problem will be one of our main themes. We shall argue that there has been a fundamental misunderstanding of how in this respect the economy in which we live actually works.

III

Though the struggle over money-wages between individuals and groups is often believed to determine

---

[1] Cf. chapter 19, Appendix.

the general level of real wages, it is, in fact, concerned with a different object. Since there is imperfect mobility of labour, and wages do not tend to an exact equality of net advantage in different occupations, any individual or group of individuals, who consent to a reduction of money-wages relatively to others, will suffer a *relative* reduction in real wages, which is a sufficient justification for them to resist it. On the other hand it would be impracticable to resist every reduction of real wages, due to a change in the purchasing-power of money which affects all workers alike; and in fact reductions of real wages arising in this way are not, as a rule, resisted unless they proceed to an extreme degree. Moreover, a resistance to reductions in money-wages applying to particular industries does not raise the same insuperable bar to an increase in aggregate employment which would result from a similar resistance to every reduction in real wages.

In other words, the struggle about money-wages primarily affects the *distribution* of the aggregate real wage between different labour-groups, and not its average amount per unit of employment, which depends, as we shall see, on a different set of forces. The effect of combination on the part of a group of workers is to protect their *relative* real wage. The *general* level of real wages depends on the other forces of the economic system.

Thus it is fortunate that the workers, though unconsciously, are instinctively more reasonable economists than the classical school, inasmuch as they resist reductions of money-wages, which are seldom or never of an all-round character, even though the existing real equivalent of these wages exceeds the marginal disutility of the existing employment; whereas they do not resist reductions of real wages, which are associated with increases in aggregate employment and leave relative money-wages unchanged, unless the reduction proceeds so far as to threaten a reduction of the real wage below the marginal disutility of the existing volume of employment. Every trade union will put up some resistance to a cut in money-wages, however small. But since no trade union would dream of striking on every occasion of a rise in the cost of living, they do not raise the obstacle to any increase in aggregate employment which is attributed to them by the classical school.

IV

We must now define the third category of unemployment, namely 'involuntary' unemployment in the strict sense, the possibility of which the classical theory does not admit.

Clearly we do not mean by 'involuntary' unemployment the mere existence of an unexhausted capacity to work. An eight-hour day does not constitute unemployment because it is not beyond human capacity to work ten hours. Nor should we regard as 'involuntary' unemployment the withdrawal of their labour by a body of workers because they do not choose to work for less than a certain real reward. Furthermore, it will be convenient to exclude 'frictional' unemployment from our definition of 'involuntary' unemployment. My definition is, therefore, as follows: *Men are involuntarily unemployed if, in the event of a small rise in the price of wage-goods relatively to the money-wage, both the aggregate supply of labour willing to work for the current money-wage and the aggregate demand for it at that wage would be greater than the existing volume of employment.* An alternative definition, which amounts, however, to the same thing, will be given in the next chapter (p. 26 below).

It follows from this definition that the equality of the real wage to the marginal disutility of employment presupposed by the second postulate, realistically interpreted, corresponds to the absence of 'involuntary' unemployment. This state of affairs we shall describe as 'full' employment, both 'frictional' and 'voluntary' unemployment being consistent with 'full' employment thus defined. This fits in, we shall find, with other characteristics of the classical theory, which is best regarded as a theory of distribution in conditions of full employment. So long as the classical postulates hold good, unemployment, which is in the above sense involuntary, cannot occur. Apparent unemployment must, therefore, be the result either of temporary loss of work of the 'between jobs' type or of intermittent demand for highly specialised resources or of the effect of a trade union 'closed shop' on the employment of free labour. Thus writers in the classical tradition, overlooking the special assumption underlying their theory, have been driven inevitably to the conclusion, perfectly logical on their assumption, that apparent unemployment (apart from the admitted exceptions) must be due at bottom to a refusal by the unemployed factors to accept a reward which corresponds to their marginal productivity. A classical economist may sympathise with labour in refusing to accept a cut in its money-wage, and he will admit that it may not be wise to make it to meet conditions which are temporary; but scientific integrity forces him to declare that this refusal is, nevertheless, at the bottom of the trouble.

Obviously, however, if the classical theory is only applicable to the case of full employment, it is fallacious to apply it to the problems of involuntary unemployment—if there be such a thing (and who will deny it?). The classical theorists resemble Euclidean geometers in a non-Euclidean world who, discovering that in experience straight lines apparently parallel often meet, rebuke the lines for not keeping straight—as the only remedy for the unfortunate collisions which are occurring. Yet, in truth, there is no remedy except to throw over the axiom of parallels and to work out a non-Euclidean geometry. Something similar is required to-day in economics. We need to throw over the second postulate of the classical doctrine and to work out the behaviour of a system in which involuntary unemployment in the strict sense is possible.

v

In emphasising our point of departure from the classical system, we must not overlook an important point of agreement. For we shall maintain the first postulate as heretofore, subject only to the same qualifications as in the classical theory; and we must pause, for a moment, to consider what this involves.

It means that, with a given organisation, equipment and technique, real wages and the volume of output (and hence of employment) are uniquely correlated, so that, in general, an increase in employment can only occur to the accompaniment of a decline in the rate of real wages. Thus I am not disputing this vital fact which the classical economists have (rightly) asserted as indefeasible. In a given state of organisation, equipment and technique, the real wage earned by a unit of labour has a unique (inverse) correlation with the volume of employment. Thus *if* employment increases, then, in the short period, the reward per unit of labour in terms of wage-goods must, in general, decline and profits increase.[1] This is simply the obverse of the familiar proposition that industry is normally working subject to decreasing returns in the short period during which equipment etc. is assumed to be constant; so that the marginal product in the wage-good industries (which governs real wages) neces-

[1] The argument runs as follows: n men are employed, the nth man adds a bushel a day to the harvest, and wages have a buying power of a bushel a day. The $n+1$th man, however, would only add ·9 bushel a day, and employment cannot, therefore, rise to $n+1$ men unless the price of corn rises relatively to wages until daily wages have a buying power of ·9 bushel. Aggregate wages would then amount to $\frac{9}{10}(n+1)$ bushels as compared with n bushels previously. Thus the employment of an additional man will, if it occurs, necessarily involve a transfer of income from those previously in work to the entrepreneurs.

sarily diminishes as employment is increased. So long, indeed, as this proposition holds, *any* means of increasing employment must lead at the same time to a diminution of the marginal product and hence of the rate of wages measured in terms of this product.

But when we have thrown over the second postulate, a decline in employment, although necessarily associated with labour's *receiving* a wage equal in value to a larger quantity of wage-goods, is not necessarily due to labour's *demanding* a larger quantity of wage-goods; and a willingness on the part of labour to accept lower money-wages is not necessarily a remedy for unemployment. The theory of wages in relation to employment, to which we are here leading up, cannot be fully elucidated, however, until chapter 19 and its Appendix have been reached.

VI

From the time of Say and Ricardo the classical economists have taught that supply creates its own demand; —meaning by this in some significant, but not clearly defined, sense that the whole of the costs of production must necessarily be spent in the aggregate, directly or indirectly, on purchasing the product.

In J. S. Mill's *Principles of Political Economy* the doctrine is expressly set forth:

> What constitutes the means of payment for commodities is simply commodities. Each person's means of paying for the productions of other people consist of those which he himself possesses. All sellers are inevitably, and by the meaning of the word, buyers. Could we suddenly double the productive powers of the country, we should double the supply of commodities in every market; but we should, by the same stroke, double the purchasing power. Everybody would bring a double demand as well as supply; everybody would be able to buy twice as much, because every one would have twice as much to offer in exchange.[1]

[1] *Principles of Political Economy*, Book III, Chap. XIV. § 2.

18

As a corollary of the same doctrine, it has been supposed that any individual act of abstaining from consumption necessarily leads to, and amounts to the same thing as, causing the labour and commodities thus released from supplying consumption to be invested in the production of capital wealth. The following passage from Marshall's *Pure Theory of Domestic Values*[1] illustrates the traditional approach:

> The whole of a man's income is expended in the purchase of services and of commodities. It is indeed commonly said that a man spends some portion of his income and saves another. But it is a familiar economic axiom that a man purchases labour and commodities with that portion of his income which he saves just as much as he does with that he is said to spend. He is said to spend when he seeks to obtain present enjoyment from the services and commodities which he purchases. He is said to save when he causes the labour and the commodities which he purchases to be devoted to the production of wealth from which he expects to derive the means of enjoyment in the future.

It is true that it would not be easy to quote comparable passages from Marshall's later work[2] or from Edgeworth or Professor Pigou. The doctrine is never stated to-day in this crude form. Nevertheless it still underlies the whole classical theory, which would collapse without it. Contemporary economists, who might hesitate to agree with Mill, do not hesitate to accept conclusions which require Mill's doctrine as their premiss. The conviction, which runs, for example, through almost all Professor Pigou's work, that money makes no real difference except frictionally and that the theory of production and employment can be

[1] P. 34.

[2] Mr J. A. Hobson, after quoting in his *Physiology of Industry* (p. 102) the above passage from Mill, points out that Marshall commented as follows on this passage as early as his *Economics of Industry*, p. 154. 'But though men have the power to purchase, they may not choose to use it.' 'But', Mr Hobson continues, 'he fails to grasp the critical importance of this fact, and appears to limit its action to periods of "crisis".' This has remained fair comment, I think, in the light of Marshall's later work.

worked out (like Mill's) as being based on 'real' exchanges with money introduced perfunctorily in a later chapter, is the modern version of the classical tradition. Contemporary thought is still deeply steeped in the notion that if people do not spend their money in one way they will spend it in another.[1] Post-war economists seldom, indeed, succeed in maintaining this standpoint *consistently*; for their thought to-day is too much permeated with the contrary tendency and with facts of experience too obviously inconsistent with their former view.[2] But they have not drawn sufficiently far-reaching consequences; and have not revised their fundamental theory.

In the first instance, these conclusions may have been applied to the kind of economy in which we actually live by false analogy from some kind of non-exchange Robinson Crusoe economy, in which the income which individuals consume or retain as a result of their productive activity is, actually and exclusively, the output *in specie* of that activity. But, apart from this, the conclusion that the *costs* of output are always covered in the aggregate by the sale-proceeds resulting from demand, has great plausibility, because it is difficult to distinguish it from another, similar-looking proposition which is indubitable, namely that the income derived in the aggregate by all the elements in the community concerned in a productive activity necessarily has a value exactly equal to the *value* of the output.

Similarly it is natural to suppose that the act of

---

[1] Cf. Alfred and Mary Marshall, *Economics of Industry*, p. 17: 'It is not good for trade to have dresses made of material which wears out quickly. For if people did not spend their means on buying new dresses they would spend them on giving employment to labour in some other way.' The reader will notice that I am again quoting from the earlier Marshall. The Marshall of the *Principles* had become sufficiently doubtful to be very cautious and evasive. But the old ideas were never repudiated or rooted out of the basic assumptions of his thought.

[2] It is the distinction of Prof. Robbins that he, almost alone, continues to maintain a consistent scheme of thought, his practical recommendations belonging to the same system as his theory.

an individual, by which he enriches himself without apparently taking anything from anyone else, must also enrich the community as a whole; so that (as in the passage just quoted from Marshall) an act of individual saving inevitably leads to a parallel act of investment. For, once more, it is indubitable that the sum of the net increments of the wealth of individuals must be exactly equal to the aggregate net increment of the wealth of the community.

Those who think in this way are deceived, nevertheless, by an optical illusion, which makes two essentially different activities appear to be the same. They are fallaciously supposing that there is a nexus which unites decisions to abstain from present consumption with decisions to provide for future consumption; whereas the motives which determine the latter are not linked in any simple way with the motives which determine the former.

It is, then, the assumption of equality between the demand price of output as a whole and its supply price which is to be regarded as the classical theory's 'axiom of parallels'. Granted this, all the rest follows —the social advantages of private and national thrift, the traditional attitude towards the rate of interest, the classical theory of unemployment, the quantity theory of money, the unqualified advantages of *laissez-faire* in respect of foreign trade and much else which we shall have to question.

VII

At different points in this chapter we have made the classical theory to depend in succession on the assumptions:

(1) that the real wage is equal to the marginal disutility of the existing employment;

(2) that there is no such thing as involuntary unemployment in the strict sense;

(3) that supply creates its own demand in the sense

that the aggregate demand price is equal to the aggregate supply price for all levels of output and employment.

These three assumptions, however, all amount to the same thing in the sense that they all stand and fall together, any one of them logically involving the other two.

Chapter 3

THE PRINCIPLE OF
EFFECTIVE DEMAND

I

We need, to start with, a few terms which will be
defined precisely later. In a given state of technique,
resources and costs, the employment of a given volume
of labour by an entrepreneur involves him in two kinds
of expense: first of all, the amounts which he pays out
to the factors of production (exclusive of other entrepreneurs) for their current services, which we shall call
the *factor cost* of the employment in question; and
secondly, the amounts which he pays out to other entrepreneurs for what he has to purchase from them
together with the sacrifice which he incurs by employing the equipment instead of leaving it idle, which we
shall call the *user cost* of the employment in question.[1]
The excess of the value of the resulting output over the
sum of its factor cost and its user cost is the profit or,
as we shall call it, the *income* of the entrepreneur. The
factor cost is, of course, the same thing, looked at from
the point of view of the entrepreneur, as what the
factors of production regard as their income. Thus the
factor cost and the entrepreneur's profit make up,
between them, what we shall define as the *total income*
resulting from the employment given by the entrepreneur. The entrepreneur's profit thus defined is, as
it should be, the quantity which he endeavours to
maximise when he is deciding what amount of employ-

[1] A precise definition of *user cost* will be given in chapter 6.

ment to offer. It is sometimes convenient, when we are looking at it from the entrepreneur's standpoint, to call the aggregate income (i.e. factor cost *plus* profit) resulting from a given amount of employment the *proceeds* of that employment. On the other hand, the aggregate supply price[1] of the output of a given amount of employment is the expectation of proceeds which will just make it worth the while of the entrepreneurs to give that employment.[2]

It follows that in a given situation of technique, resources and factor cost per unit of employment, the amount of employment, both in each individual firm and industry and in the aggregate, depends on the amount of the proceeds which the entrepreneurs expect to receive from the corresponding output.[3] For entrepreneurs will endeavour to fix the amount of employ-

---

[1] Not to be confused (*vide infra*) with the supply price of a unit of output in the ordinary sense of this term.

[2] The reader will observe that I am deducting the user cost both from the *proceeds* and from the *aggregate supply price* of a given volume of output, so that both these terms are to be interpreted *net* of user cost; whereas the aggregate sums paid by the purchasers are, of course, *gross* of user cost. The reasons why this is convenient will be given in chapter 6. The essential point is that the aggregate proceeds and aggregate supply price net of user cost can be defined uniquely and unambiguously; whereas, since user cost is obviously dependent both on the degree of integration of industry and on the extent to which entrepreneurs buy from one another, there can be no definition of the aggregate sums paid by purchasers, *inclusive* of user cost, which is independent of these factors. There is a similar difficulty even in defining supply price in the ordinary sense for an individual producer; and in the case of the aggregate supply price of *output as a whole* serious difficulties of duplication are involved, which have not always been faced. If the term is to be interpreted gross of user cost, they can only be overcome by making special assumptions relating to the integration of entrepreneurs in groups according as they produce consumption-goods or capital-goods, which are obscure and complicated in themselves and do not correspond to the facts. If, however, aggregate supply price is defined as above *net* of user cost, these difficulties do not arise. The reader is advised, however, to await the fuller discussion in chapter 6 and its Appendix.

[3] An entrepreneur, who has to reach a practical decision as to his scale of production, does not, of course, entertain a single undoubting expectation of what the sale-proceeds of a given output will be, but several hypothetical expectations held with varying degrees of probability and definiteness. By his expectation of proceeds I mean, therefore, that expectation of proceeds which, if it were held with certainty, would lead to the same behaviour as does the bundle of vague and more various possibilities which actually makes up his state of expectation when he reaches his decision.

ment at the level which they expect to maximise the excess of the proceeds over the factor cost.

Let Z be the aggregate supply price of the output from employing N men, the relationship between Z and N being written $Z = \phi(N)$, which can be called the *aggregate supply function*.[1] Similarly, let D be the proceeds which entrepreneurs expect to receive from the employment of N men, the relationship between D and N being written $D = f(N)$, which can be called the *aggregate demand function*.

Now if for a given value of N the expected proceeds are greater than the aggregate supply price, i.e. if D is greater than Z, there will be an incentive to entrepreneurs to increase employment beyond N and, if necessary, to raise costs by competing with one another for the factors of production, up to the value of N for which Z has become equal to D. Thus the volume of employment is given by the point of intersection between the aggregate demand function and the aggregate supply function; for it is at this point that the entrepreneurs' expectation of profits will be maximised. The value of D at the point of the aggregate demand function, where it is intersected by the aggregate supply function, will be called *the effective demand*. Since this is the substance of the General Theory of Employment, which it will be our object to expound, the succeeding chapters will be largely occupied with examining the various factors upon which these two functions depend.

The classical doctrine, on the other hand, which used to be expressed categorically in the statement that 'Supply creates its own Demand' and continues to underlie all orthodox economic theory, involves a special assumption as to the relationship between these two functions. For 'Supply creates its own Demand' must mean that $f(N)$ and $\phi(N)$ are equal for *all* values

---

[1] In chapter 20 a function closely related to the above will be called the employment function.

of N, i.e. for all levels of output and employment; and that when there is an increase in $Z(=\phi(N))$ corresponding to an increase in N, $D(=f(N))$ necessarily increases by the same amount as Z. The classical theory assumes, in other words, that the aggregate demand price (or proceeds) always accommodates itself to the aggregate supply price; so that, whatever the value of N may be, the proceeds D assume a value equal to the aggregate supply price Z which corresponds to N. That is to say, effective demand, instead of having a unique equilibrium value, is an infinite range of values all equally admissible; and the amount of employment is indeterminate except in so far as the marginal disutility of labour sets an upper limit.

If this were true, competition between entrepreneurs would always lead to an expansion of employment up to the point at which the supply of output as a whole ceases to be elastic, i.e. where a further increase in the value of the effective demand will no longer be accompanied by any increase in output. Evidently this amounts to the same thing as full employment. In the previous chapter we have given a definition of full employment in terms of the behaviour of labour. An alternative, though equivalent, criterion is that at which we have now arrived, namely a situation in which aggregate employment is inelastic in response to an increase in the effective demand for its output. Thus Say's law, that the aggregate demand price of output as a whole is equal to its aggregate supply price for all volumes of output, is equivalent to the proposition that there is no obstacle to full employment. If, however, this is not the true law relating the aggregate demand and supply functions, there is a vitally important chapter of economic theory which remains to be written and without which all discussions concerning the volume of aggregate employment are futile.

II

A brief summary of the theory of employment to be worked out in the course of the following chapters may, perhaps, help the reader at this stage, even though it may not be fully intelligible. The terms involved will be more carefully defined in due course. In this summary we shall assume that the money-wage and other factor costs are constant per unit of labour employed. But this simplification, with which we shall dispense later, is introduced solely to facilitate the exposition. The essential character of the argument is precisely the same whether or not money-wages, etc., are liable to change.

The outline of our theory can be expressed as follows. When employment increases, aggregate real income is increased. The psychology of the community is such that when aggregate real income is increased aggregate consumption is increased, but not by so much as income. Hence employers would make a loss if the whole of the increased employment were to be devoted to satisfying the increased demand for immediate consumption. Thus, to justify any given amount of employment there must be an amount of current investment sufficient to absorb the excess of total output over what the community chooses to consume when employment is at the given level. For unless there is this amount of investment, the receipts of the entrepreneurs will be less than is required to induce them to offer the given amount of employment. It follows, therefore, that, given what we shall call the community's propensity to consume, the equilibrium level of employment, i.e. the level at which there is no inducement to employers as a whole either to expand or to contract employment, will depend on the amount of current investment. The amount of current investment will depend, in turn, on what we shall call the inducement to invest; and the inducement to invest will be found to depend on the relation between the schedule of the marginal efficiency of capital and the complex of rates of interest on loans of various maturities and risks.

Thus, given the propensity to consume and the rate of new investment, there will be only one level of employment consistent with equilibrium; since any other level will lead to inequality between the aggregate supply price of output as a whole and its aggregate demand price. This level cannot be *greater* than full employment, i.e. the real wage cannot be less than the marginal disutility of labour. But there is no reason in general for expecting it to be *equal* to full employment. The effective demand associated with full employment is a special case, only realised when the propensity to consume and the inducement to invest stand in a particular relationship to one another. This particular relationship, which corresponds to the assumptions of the classical theory, is in a sense an optimum relationship. But it can only exist when, by accident or design, current investment provides an amount of demand just equal to the excess of the aggregate supply price of the output resulting from full employment over what the community will choose to spend on consumption when it is fully employed.

This theory can be summed up in the following propositions:

(1) In a given situation of technique, resources and costs, income (both money-income and real income) depends on the volume of employment N.

(2) The relationship between the community's income and what it can be expected to spend on consumption, designated by D_1, will depend on the psychological characteristic of the community, which we shall call its *propensity to consume*. That is to say, consumption will depend on the level of aggregate income and, therefore, on the level of employment N, except when there is some change in the propensity to consume.

(3) The amount of labour N which the entrepreneurs decide to employ depends on the sum (D) of *two* quantities, namely D_1, the amount which the community is expected to spend on consumption, and D_2, the amount which it is expected to devote to new investment. D is what we have called above the *effective demand*.

(4) Since $D_1 + D_2 = D = \phi(N)$, where ϕ is the aggregate supply function, and since, as we have seen in (2) above, D_1 is a function of N, which we may write $\chi(N)$, depending on the propensity to consume, it follows that $\phi(N) - \chi(N) = D_2$.

(5) Hence the volume of employment in equilibrium depends on (i) the aggregate supply function, ϕ, (ii) the propensity to consume, χ, and (iii) the volume of investment, D_2. This is the essence of the General Theory of Employment.

(6) For every value of N there is a corresponding marginal productivity of labour in the wage-goods industries; and it is this which determines the real wage. (5) is, therefore, subject to the condition that N cannot *exceed* the value which reduces the real wage to equality with the marginal disutility of labour. This means that not all changes in D are compatible with our temporary assumption that money-wages are constant. Thus it will be essential to a full statement of our theory to dispense with this assumption.

(7) On the classical theory, according to which $D = \phi(N)$ for *all* values of N, the volume of employment is in neutral equilibrium for all values of N less than its maximum value; so that the forces of competition between entrepreneurs may be expected to push it to this maximum value. Only at this point, on the classical theory, can there be stable equilibrium.

(8) *When employment increases, D_1 will increase, but not by so much as D;* since when our income increases our consumption increases also, but not by so much. The key to our practical problem is to be found in this

psychological law. For it follows from this that the greater the volume of employment the greater will be the gap between the aggregate supply price (Z) of the corresponding output and the sum (D_1) which the entrepreneurs can expect to get back out of the expenditure of consumers. Hence, if there is no change in the propensity to consume, employment cannot increase, unless at the same time D_2 is increasing so as to fill the increasing gap between Z and D_1. Thus—except on the special assumptions of the classical theory according to which there is some force in operation which, when employment increases, always causes D_2 to increase sufficiently to fill the widening gap between Z and D_1—the economic system may find itself in stable equilibrium with N at a level below full employment, namely at the level given by the intersection of the aggregate demand function with the aggregate supply function.

Thus the volume of employment is not determined by the marginal disutility of labour measured in terms of real wages, except in so far as the supply of labour available at a given real wage sets a *maximum* level to employment. The propensity to consume and the rate of new investment determine between them the volume of employment, and the volume of employment is uniquely related to a given level of real wages—not the other way round. If the propensity to consume and the rate of new investment result in a deficient effective demand, the actual level of employment will fall short of the supply of labour potentially available at the existing real wage, and the equilibrium real wage will be *greater* than the marginal disutility of the equilibrium level of employment.

This analysis supplies us with an explanation of the paradox of poverty in the midst of plenty. For the mere existence of an insufficiency of effective demand may, and often will, bring the increase of employment to a standstill *before* a level of full employment has been reached. The insufficiency of effective demand will inhibit the process of production in spite of the fact that the marginal product of labour still exceeds in value the marginal disutility of employment.

Moreover the richer the community, the wider will tend to be the gap between its actual and its potential production; and therefore the more obvious and outrageous the defects of the economic system. For a poor community will be prone to consume by far the greater part of its output, so that a very modest measure of investment will be sufficient to provide full employment; whereas a wealthy community will have to discover much ampler opportunities for investment if the saving propensities of its wealthier members are to be compatible with the employment of its poorer members. If in a potentially wealthy community the inducement to invest is weak, then, in spite of its potential wealth, the working of the principle of effective demand will compel it to reduce its actual output, until, in spite of its potential wealth, it has become so poor that its surplus over its consumption is sufficiently diminished to correspond to the weakness of the inducement to invest.

But worse still. Not only is the marginal propensity to consume[1] weaker in a wealthy community, but, owing to its accumulation of capital being already larger, the opportunities for further investment are less attractive unless the rate of interest falls at a sufficiently rapid rate; which brings us to the theory of the rate of interest and to the reasons why it does not automatically fall to the appropriate level, which will occupy Book IV.

Thus the analysis of the propensity to consume, the definition of the marginal efficiency of capital and the theory of the rate of interest are the three main gaps in our existing knowledge which it will be necessary to fill. When this has been accomplished,

---

[1] Defined in chapter 10, below.

we shall find that the theory of prices falls into its proper place as a matter which is subsidiary to our general theory. We shall discover, however, that money plays an essential part in our theory of the rate of interest; and we shall attempt to disentangle the peculiar characteristics of money which distinguish it from other things.

III

The idea that we can safely neglect the aggregate demand function is fundamental to the Ricardian economics, which underlie what we have been taught for more than a century. Malthus, indeed, had vehemently opposed Ricardo's doctrine that it was impossible for effective demand to be deficient; but vainly. For, since Malthus was unable to explain clearly (apart from an appeal to the facts of common observation) how and why effective demand could be deficient or excessive, he failed to furnish an alternative construction; and Ricardo conquered England as completely as the Holy Inquisition conquered Spain. Not only was his theory accepted by the city, by statesmen and by the academic world. But controversy ceased; the other point of view completely disappeared; it ceased to be discussed. The great puzzle of effective demand with which Malthus had wrestled vanished from economic literature. You will not find it mentioned even once in the whole works of Marshall, Edgeworth and Professor Pigou, from whose hands the classical theory has received its most mature embodiment. It could only live on furtively, below the surface, in the underworlds of Karl Marx, Silvio Gesell or Major Douglas.

The completeness of the Ricardian victory is something of a curiosity and a mystery. It must have been due to a complex of suitabilities in the doctrine to the environment into which it was projected. That it reached conclusions quite different from what the ordinary uninstructed person would expect, added, I suppose, to its intellectual prestige. That its teaching, translated into practice, was austere and often unpalatable, lent it virtue. That it was adapted to carry a vast and consistent logical superstructure, gave it beauty. That it could explain much social injustice and apparent cruelty as an inevitable incident in the scheme of progress, and the attempt to change such things as likely on the whole to do more harm than good, commended it to authority. That it afforded a measure of justification to the free activities of the individual capitalist, attracted to it the support of the dominant social force behind authority.

But although the doctrine itself has remained unquestioned by orthodox economists up to a late date, its signal failure for purposes of scientific prediction has greatly impaired, in the course of time, the prestige of its practitioners. For professional economists, after Malthus, were apparently unmoved by the lack of correspondence between the results of their theory and the facts of observation;—a discrepancy which the ordinary man has not failed to observe, with the result of his growing unwillingness to accord to economists that measure of respect which he gives to other groups of scientists whose theoretical results are confirmed by observation when they are applied to the facts.

The celebrated *optimism* of traditional economic theory, which has led to economists being looked upon as Candides, who, having left this world for the cultivation of their gardens, teach that all is for the best in the best of all possible worlds provided we will let well alone, is also to be traced, I think, to their having neglected to take account of the drag on prosperity which can be exercised by an insufficiency of effective demand. For there would obviously be a natural tendency towards the optimum employment of resources in a society which was functioning after the manner of the classical postulates. It may well be that the classical theory represents the way in which we should like our economy to behave. But to assume that it actually does so is to assume our difficulties away.

BOOK II
DEFINITIONS AND IDEAS

Chapter 4

THE CHOICE OF UNITS

In this and the next three chapters we shall be occupied with an attempt to clear up certain perplexities which have no peculiar or exclusive relevance to the problems which it is our special purpose to examine. Thus these chapters are in the nature of a digression, which will prevent us for a time from pursuing our main theme. Their subject-matter is only discussed here because it does not happen to have been already treated elsewhere in a way which I find adequate to the needs of my own particular enquiry.

The three perplexities which most impeded my progress in writing this book, so that I could not express myself conveniently until I had found some solution for them, are: firstly, the choice of the units of quantity appropriate to the problems of the economic system as a whole; secondly, the part played by expectation in economic analysis; and, thirdly, the definition of income.

II

That the units, in terms of which economists commonly work, are unsatisfactory can be illustrated by the concepts of the national dividend, the stock of real capital and the general price-level:

(i) The national dividend, as defined by Marshall

and Professor Pigou,[1] measures the volume of current output or real income and not the value of output or money-income.[2] Furthermore, it depends, in some sense, on *net* output;—on the net addition, that is to say, to the resources of the community available for consumption or for retention as capital stock, due to the economic activities and sacrifices of the current period, after allowing for the wastage of the stock of real capital existing at the commencement of the period. On this basis an attempt is made to erect a quantitative science. But it is a grave objection to this definition for such a purpose that the community's output of goods and services is a non-homogeneous complex which cannot be measured, strictly speaking, except in certain special cases, as for example when all the items of one output are included in the same proportions in another output.

(ii) The difficulty is even greater when, in order to calculate net output, we try to measure the net addition to capital equipment; for we have to find some basis for a quantitative comparison between the new items of equipment produced during the period and the old items which have perished by wastage. In order to arrive at the net national dividend, Professor Pigou[3] deducts such obsolescence, etc., 'as may fairly be called "normal"; and the practical test of normality is that the depletion is sufficiently regular to be foreseen, if not in detail, at least in the large'. But, since this deduction is not a deduction in terms of money, he is involved in assuming that there can be a change in physical quantity, although there has been no physical change; i.e. he is covertly introducing changes in *value*.

---

[1] *Vide* Pigou, *Economics of Welfare, passim,* and particularly Part I, Chap. iii.

[2] Though, as a convenient compromise, the real income, which is taken to constitute the national dividend, is usually limited to those goods and services which can be bought for money.

[3] *Economics of Welfare,* Part I, Chap. v, on 'What is Meant by Maintaining Capital Intact'; as amended by a recent article in the *Economic Journal,* June 1935, p. 225.

Moreover, he is unable to devise any satisfactory formula[1] to evaluate new equipment against old when, owing to changes in technique, the two are not identical. I believe that the concept at which Professor Pigou is aiming is the right and appropriate concept for economic analysis. But, until a satisfactory system of units has been adopted, its precise definition is an impossible task. The problem of comparing one real output with another and of then calculating net output by setting off new items of equipment against the wastage of old items presents conundrums which permit, one can confidently say, of no solution.

(iii) Thirdly, the well-known, but unavoidable, element of vagueness which admittedly attends the concept of the general price-level makes this term very unsatisfactory for the purposes of a causal analysis, which ought to be exact.

Nevertheless these difficulties are rightly regarded as 'conundrums'. They are 'purely theoretical' in the sense that they never perplex, or indeed enter in any way into, business decisions and have no relevance to the causal sequence of economic events, which are clear-cut and determinate in spite of the quantitative indeterminacy of these concepts. It is natural, therefore, to conclude that they not only lack precision but are unnecessary. Obviously our quantitative analysis must be expressed without using any quantitatively vague expressions. And, indeed, as soon as one makes the attempt, it becomes clear, as I hope to show, that one can get on much better without them.

The fact that two incommensurable collections of miscellaneous objects cannot in themselves provide the material for a quantitative analysis need not, of course, prevent us from making approximate statistical comparisons, depending on some broad element of judgment rather than of strict calculation, which may possess significance and validity within certain limits.

[1] Cf. Prof. Hayek's criticisms, *Economica*, August 1935, p. 247.

But the proper place for such things as net real output and the general level of prices lies within the field of historical and statistical description, and their purpose should be to satisfy historical or social curiosity, a purpose for which perfect precision—such as our causal analysis requires, whether or not our knowledge of the actual values of the relevant quantities is complete or exact—is neither usual nor necessary. To say that net output to-day is greater, but the price-level lower, than ten years ago or one year ago, is a proposition of a similar character to the statement that Queen Victoria was a better queen but not a happier woman than Queen Elizabeth—a proposition not without meaning and not without interest, but unsuitable as material for the differential calculus. Our precision will be a mock precision if we try to use such partly vague and non-quantitative concepts as the basis of a quantitative analysis.

III

On *every* particular occasion, let it be remembered, an entrepreneur is concerned with decisions as to the scale on which to work a given capital equipment; and when we say that the expectation of an increased demand, i.e. a raising of the aggregate demand function, will lead to an increase in aggregate output, we really mean that the firms, which own the capital equipment, will be induced to associate with it a greater aggregate employment of labour. In the case of an individual firm or industry producing a homogeneous product we can speak legitimately, if we wish, of increases or decreases of output. But when we are aggregating the activities of all firms, we cannot speak accurately except in terms of quantities of employment applied to a given equipment. The concepts of output as a whole and its price-level are not required in this context, since we have no need of an absolute measure of current aggregate output, such as would enable us to compare its amount with the amount which would result from the association of a different capital equipment with a different quantity of employment. When, for purposes of description or rough comparison, we wish to speak of an increase of output, we must rely on the general presumption that the amount of employment associated with a given capital equipment will be a satisfactory index of the amount of resultant output;—the two being presumed to increase and decrease together, though not in a definite numerical proportion.

In dealing with the theory of employment I propose, therefore, to make use of only two fundamental units of quantity, namely, quantities of money-value and quantities of employment. The first of these is strictly homogeneous, and the second can be made so. For, in so far as different grades and kinds of labour and salaried assistance enjoy a more or less fixed relative remuneration, the quantity of employment can be sufficiently defined for our purpose by taking an hour's employment of ordinary labour as our unit and weighting an hour's employment of special labour in proportion to its remuneration; i.e. an hour of special labour remunerated at double ordinary rates will count as two units. We shall call the unit in which the quantity of employment is measured the labour-unit; and the money-wage of a labour-unit we shall call the wage-unit.[1] Thus, if E is the wages (and salaries) bill, W the wage-unit, and N the quantity of employment, $E = N.W$.

This assumption of homogeneity in the supply of labour is not upset by the obvious fact of great differences in the specialised skill of individual workers and in their suitability for different occupations. For,

[1] If X stands for any quantity measured in terms of money, it will often be convenient to write X_w for the same quantity measured in terms of the wage-unit.

if the remuneration of the workers is proportional to their efficiency, the differences are dealt with by our having regarded individuals as contributing to the supply of labour in proportion to their remuneration; whilst if, as output increases, a given firm has to bring in labour which is less and less efficient for its special purposes per wage-unit paid to it, this is merely one factor among others leading to a diminishing return from the capital equipment in terms of output as more labour is employed on it. We subsume, so to speak, the non-homogeneity of equally remunerated labour units in the equipment, which we regard as less and less adapted to employ the available labour units as output increases, instead of regarding the available labour units as less and less adapted to use a homogeneous capital equipment. Thus if there is no surplus of specialised or practised labour and the use of less suitable labour involves a higher labour cost per unit of output, this means that the rate at which the return from the equipment diminishes as employment increases is more rapid than it would be if there were such a surplus.[1] Even in the limiting case where different labour units were so highly specialised as to be altogether incapable of being substituted for one another, there is no awkwardness; for this merely means that the elasticity of supply of output from a particular type of capital equipment falls suddenly to zero when all the available labour specialised to its use is already employed.[2] Thus our assumption of a homo-

---

[1] This is the main reason why the supply price of output rises with increasing demand even when there is still a surplus of equipment identical in type with the equipment in use. If we suppose that the surplus supply of labour forms a pool equally available to all entrepreneurs and that labour employed for a given purpose is rewarded, in part at least, per unit of effort and not with strict regard to its efficiency in its actual particular employment (which is in most cases the realistic assumption to make), the diminishing efficiency of the labour employed is an outstanding example of rising supply price with increasing output, not due to internal diseconomies.

[2] How the supply curve in ordinary use is supposed to deal with the above difficulty I cannot say, since those who use this curve have not made their assumptions very clear. Probably they are assuming that labour

geneous unit of labour involves no difficulties unless there is great instability in the relative remuneration of different labour-units; and even this difficulty can be dealt with, if it arises, by supposing a rapid liability to change in the supply of labour and the shape of the aggregate supply function.

It is my belief that much unnecessary perplexity can be avoided if we limit ourselves strictly to the two units, money and labour, when we are dealing with the behaviour of the economic system as a whole; reserving the use of units of particular outputs and equipments to the occasions when we are analysing the output of individual firms or industries in isolation; and the use of vague concepts, such as the quantity of output as a whole, the quantity of capital equipment as a whole and the general level of prices, to the occasions when we are attempting some historical comparison which is within certain (perhaps fairly wide) limits avowedly unprecise and approximate.

employed for a given purpose is always rewarded with strict regard to its efficiency for that purpose. But this is unrealistic. Perhaps the essential reason for treating the varying efficiency of labour as though it belonged to the equipment lies in the fact that the increasing surpluses, which emerge as output is increased, accrue in practice mainly to the owners of the equipment and not to the more efficient workers (though these may get an advantage through being employed more regularly and by receiving earlier promotion); that is to say, men of differing efficiency working at the same job are seldom paid at rates closely proportional to their efficiencies. Where, however, increased pay for higher efficiency occurs, and in so far as it occurs my method takes account of it; since in calculating the number of labour units employed, the individual workers are weighted in proportion to their remuneration. On my assumptions interesting complications obviously arise where we are dealing with particular supply curves since their shape will depend on the demand for suitable labour in other directions. To ignore these complications would, as I have said, be unrealistic. But we need not consider them when we are dealing with employment as a whole, provided we assume that a given volume of effective demand has a particular distribution of this demand between different products uniquely associated with it. It may be, however, that this would not hold good irrespective of the particular cause of the change in demand. E.g. an increase in effective demand due to an increased propensity to consume might find itself faced by a different aggregate supply function from that which would face an equal increase in demand due to an increased inducement to invest. All this, however, belongs to the detailed analysis of the general ideas here set forth, which it is no part of my immediate purpose to pursue.

43

It follows that we shall measure changes in current output by reference to the number of hours of labour paid for (whether to satisfy consumers or to produce fresh capital equipment) on the existing capital equipment, hours of skilled labour being weighted in proportion to their remuneration. We have no need of a quantitative comparison between this output and the output which would result from associating a different set of workers with a different capital equipment. To predict how entrepreneurs possessing a given equipment will respond to a shift in the aggregate demand function it is not necessary to know how the quantity of the resulting output, the standard of life and the general level of prices would compare with what they were at a different date or in another country.

IV

It is easily shown that the conditions of supply, such as are usually expressed in terms of the supply curve, and the elasticity of supply relating output to price, can be handled in terms of our two chosen units by means of the aggregate supply function, without reference to quantities of output, whether we are concerned with a particular firm or industry or with economic activity as a whole. For the aggregate supply function for a given firm (and similarly for a given industry or for industry as a whole) is given by

$$Z_r = \phi_r(N_r),$$

where Z_r is the proceeds (net of user cost) the expectation of which will induce a level of employment N_r. If, therefore, the relation between employment and output is such that an employment N_r results in an output O_r, where $O_r = \psi_r(N_r)$, it follows that

$$p = \frac{Z_r + U_r(N_r)}{O_r.} = \frac{\phi_r(N_r) + U_r(N_r)}{\psi_r(N_r)}$$

is the ordinary supply curve, where $U_r(N_r)$ is the (expected) user cost corresponding to a level of employment N_r.

Thus in the case of each homogeneous commodity, for which $O_r = \psi_r(N_r)$ has a definite meaning, we can evaluate $Z_r = \phi_r(N_r)$ in the ordinary way; but we can then aggregate the N_r's in a way in which we cannot aggregate the O_r's, since ΣO_r is not a numerical quantity. Moreover, if we can assume that, in a given environment, a given aggregate employment will be distributed in a unique way between different industries, so that N_r is a function of N, further simplifications are possible.

Chapter 5

EXPECTATION AS DETERMINING OUTPUT AND EMPLOYMENT

I

All production is for the purpose of ultimately satisfying a consumer. Time usually elapses, however—and sometimes much time—between the incurring of costs by the producer (with the consumer in view) and the purchase of the output by the ultimate consumer. Meanwhile the entrepreneur (including both the producer and the investor in this description) has to form the best expectations[1] he can as to what the consumers will be prepared to pay when he is ready to supply them (directly or indirectly) after the elapse of what may be a lengthy period; and he has no choice but to be guided by these expectations, if he is to produce at all by processes which occupy time.

These expectations, upon which business decisions depend, fall into two groups, certain individuals or firms being specialised in the business of framing the first type of expectation and others in the business of framing the second. The first type is concerned with the price which a manufacturer can expect to get for his 'finished' output at the time when he commits himself to starting the process which will produce it; output being 'finished' (from the point of view of the manufacturer) when it is ready to be used or to be sold to a second party. The

---

[1] For the method of arriving at an equivalent of these expectations in terms of sale-proceeds see footnote 3 to p. 24 above.

46

second type is concerned with what the entrepreneur can hope to earn in the shape of future returns if he purchases (or, perhaps, manufactures) 'finished' output as an addition to his capital equipment. We may call the former *short-term expectation* and the latter *long-term expectation*.

Thus the behaviour of each individual firm in deciding its daily[1] output will be determined by its *short-term expectations*—expectations as to the cost of output on various possible scales and expectations as to the sale-proceeds of this output; though, in the case of additions to capital equipment and even of sales to distributors, these short-term expectations will largely depend on the long-term (or medium-term) expectations of other parties. It is upon these various expectations that the amount of employment which the firms offer will depend. The *actually realised* results of the production and sale of output will only be relevant to employment in so far as they cause a modification of subsequent expectations. Nor, on the other hand, are the original expectations relevant, which led the firm to acquire the capital equipment and the stock of intermediate products and half-finished materials with which it finds itself at the time when it has to decide the next day's output. Thus, on each and every occasion of such a decision, the decision will be made, with reference indeed to this equipment and stock, but in the light of the *current* expectations of *prospective* costs and sale-proceeds.

Now, in general, a *change* in expectations (whether short-term or long-term) will only produce its full effect on employment over a considerable period. The change in employment due to a change in expectations will not be the same on the second day after the change as on the first, or the same on the

---

[1] *Daily* here stands for the shortest interval after which the firm is free to revise its decision as to how much employment to offer. It is, so to speak, the minimum effective unit of economic time.

third day as on the second, and so on, even though there be no further change in expectations. In the case of short-term expectations this is because changes in expectation are not, as a rule, sufficiently violent or rapid, when they are for the worse, to cause the abandonment of work on all the productive processes which, in the light of the revised expectation, it was a mistake to have begun; whilst, when they are for the better, some time for preparation must needs elapse before employment can reach the level at which it would have stood if the state of expectation had been revised sooner. In the case of long-term expectations, equipment which will not be replaced will continue to give employment until it is worn out; whilst when the change in long-term expectations is for the better, employment may be at a higher level at first, than it will be after there has been time to adjust the equipment to the new situation.

If we suppose a state of expectation to continue for a sufficient length of time for the effect on employment to have worked itself out so completely that there is, broadly speaking, no piece of employment going on which would not have taken place if the new state of expectation had always existed, the steady level of employment thus attained may be called the long-period employment[1] corresponding to that state of expectation. It follows that, although expectation may change so frequently that the actual level of employment has never had time to reach the long-period employment corresponding to the existing state of expectation, nevertheless every state of expectation has its definite corresponding level of long-period employment.

Let us consider, first of all, the process of transition

---

[1] It is not necessary that the level of long-period employment should be *constant*, i.e. long-period conditions are not necessarily static. For example, a steady increase in wealth or population may constitute a part of the unchanging expectation. The only condition is that the existing expectations should have been foreseen sufficiently far ahead.

to a long-period position due to a change in expectation, which is not confused or interrupted by any further change in expectation. We will first suppose that the change is of such a character that the new long-period employment will be greater than the old. Now, as a rule, it will only be the rate of input which will be much affected at the beginning, that is to say, the volume of work on the earlier stages of new processes of production, whilst the output of consumption-goods and the amount of employment on the later stages of processes which were started before the change will remain much the same as before. In so far as there were stocks of partly finished goods, this conclusion may be modified; though it is likely to remain true that the initial increase in employment will be modest. As, however, the days pass by, employment will gradually increase. Moreover, it is easy to conceive of conditions which will cause it to increase at some stage to a *higher* level than the new long-period employment. For the process of building up capital to satisfy the new state of expectation may lead to more employment and also to more current consumption than will occur when the long-period position has been reached. Thus the change in expectation may lead to a gradual crescendo in the level of employment, rising to a peak and then declining to the new long-period level. The same thing may occur even if the new long-period level is the *same* as the old, if the change represents a change in the direction of consumption which renders certain existing processes and their equipment obsolete. Or again, if the new long-period employment is less than the old, the level of employment during the transition may fall for a time *below* what the new long-period level is going to be. Thus a mere change in expectation is capable of producing an oscillation of the same kind of shape as a cyclical movement, in the course of working itself out. It was movements of this kind which I discussed in my *Treatise on Money* in connection with the building up or the depletion of stocks of working and liquid capital consequent on change.

An uninterrupted process of transition, such as the above, to a new long-period position can be complicated in detail. But the actual course of events is more complicated still. For the state of expectation is liable to constant change, a new expectation being superimposed long before the previous change has fully worked itself out; so that the economic machine is occupied at any given time with a number of overlapping activities, the existence of which is due to various past states of expectation.

II

This leads us to the relevance of this discussion for our present purpose. It is evident from the above that the level of employment at any time depends, in a sense, not merely on the existing state of expectation but on the states of expectation which have existed over a certain past period. Nevertheless past expectations, which have not yet worked themselves out, are embodied in the to-day's capital equipment with reference to which the entrepreneur has to make to-day's decisions, and only influence his decisions in so far as they are so embodied. It follows, therefore, that, in spite of the above, to-day's employment can be correctly described as being governed by to-day's expectations taken in conjunction with to-day's capital equipment.

Express reference to current long-term expectations can seldom be avoided. But it will often be safe to omit express reference to *short-term* expectation, in view of the fact that in practice the process of revision of short-term expectation is a gradual and continuous one, carried on largely in the light of realised results; so that expected and realised results run into and overlap one another in their influence. For, although output and employment are determined by the producer's short-term expectations and not by past results, the most recent results usually play a predominant part in determining what these expectations are. It would be too complicated to work out the expectations *de novo* whenever a productive process was being started; and it would, moreover, be a waste of time since a large part of the circumstances usually continue substantially unchanged from one day to the next. Accordingly it is sensible for producers to base their expectations on the assumption that the most recently realised results will continue, except in so far as there are definite reasons for expecting a change. Thus in practice there is a large overlap between the effects on employment of the realised sale-proceeds of recent output and those of the sale-proceeds expected from current input; and producers' forecasts are more often gradually modified in the light of results than in anticipation of prospective changes.[1]

Nevertheless, we must not forget that, in the case of durable goods, the producer's short-term expectations are based on the current long-term expectations of the investor; and it is of the nature of long-term expectations that they cannot be checked at short intervals in the light of realised results. Moreover, as we shall see in chapter 12, where we shall consider long-term expectations in more detail, they are liable to sudden revision. Thus the factor of current long-term expectations cannot be even approximately eliminated or replaced by realised results.

[1] This emphasis on the expectation entertained when the decision to produce is taken, meets, I think, Mr Hawtrey's point that input and employment are influenced by the accumulation of stocks *before* prices have fallen or disappointment in respect of output is reflected in a realised loss relatively to expectation. For the accumulation of unsold stocks (or decline of forward orders) is precisely the kind of event which is most likely to cause input to differ from what the mere statistics of the sale-proceeds of previous output would indicate if they were to be projected without criticism into the next period.

Chapter 6

THE DEFINITION OF INCOME, SAVING AND INVESTMENT

I. *Income*

During any period of time an entrepreneur will have sold finished output to consumers or to other entrepreneurs for a certain sum which we will designate as A. He will also have spent a certain sum, designated by A_1, on purchasing finished output from other entrepreneurs. And he will end up with a capital equipment, which term includes both his stocks of unfinished goods or working capital and his stocks of finished goods, having a value G.

Some part, however, of $A + G - A_1$ will be attributable, not to the activities of the period in question, but to the capital equipment which he had at the beginning of the period. We must, therefore, in order to arrive at what we mean by the *income* of the current period, deduct from $A + G - A_1$ a certain sum, to represent that part of its value which has been (in some sense) contributed by the equipment inherited from the previous period. The problem of defining income is solved as soon as we have found a satisfactory method for calculating this deduction.

There are two possible principles for calculating it, each of which has a certain significance;—one of them in connection with production, and the other in connection with consumption. Let us consider them in turn.

(i) The actual value G of the capital equipment at

the end of the period is the net result of the entrepreneur, on the one hand, having maintained and improved it during the period, both by purchases from
other entrepreneurs and by work done upon it by
himself, and, on the other hand, having exhausted or
depreciated it through using it to produce output. If
he had decided *not* to use it to produce output, there is,
nevertheless, a certain optimum sum which it would
have paid him to spend on maintaining and improving
it. Let us suppose that, in this event, he would have
spent B' on its maintenance and improvement, and
that, having had this spent on it, it would have been
worth G' at the end of the period. That is to say,
$G' - B'$ is the maximum net value which might have
been conserved from the previous period, if it had not
been used to produce A. The excess of this potential
value of the equipment over $G - A_1$ is the measure of
what has been sacrificed (one way or another) to produce A. Let us call this quantity, namely

$$(G' - B') - (G - A_1),$$

which measures the sacrifice of value involved in the
production of A, the *user cost* of A. *User cost* will be
written U.[1] The amount paid out by the entrepreneur
to the other factors of production in return for their
services, which from their point of view is their income,
we will call the *factor cost* of A. The sum of the factor
cost F and the user cost U we shall call the *prime cost*
of the output A.

We can then define the *income*[2] of the entrepreneur as
being the excess of the value of his finished output sold
during the period over his prime cost. The entrepreneur's income, that is to say, is taken as being equal
to the quantity, depending on his scale of production,
which he endeavours to maximise, i.e. to his gross profit

[1] Some further observations on user cost are given in an Appendix to this
chapter.
[2] As distinguished from his *net income* which we shall define below.

in the ordinary sense of this term;—which agrees with common sense. Hence, since the income of the rest of the community is equal to the entrepreneur's factor cost, aggregate income is equal to $A - U$.

Income, thus defined, is a completely unambiguous quantity. Moreover, since it is the entrepreneur's expectation of the excess of this quantity over his outgoings to the other factors of production which he endeavours to maximise when he decides how much employment to give to the other factors of production, it is the quantity which is causally significant for employment.

It is conceivable, of course, that $G - A_1$ may exceed $G' - B'$, so that user cost will be negative. For example, this may well be the case if we happen to choose our period in such a way that input has been increasing during the period but without there having been time for the increased output to reach the stage of being finished and sold. It will also be the case, whenever there is positive investment, if we imagine industry to be so much integrated that entrepreneurs make most of their equipment for themselves. Since, however, user cost is only negative when the entrepreneur has been increasing his capital equipment by his own labour, we can, in an economy where capital equipment is largely manufactured by different firms from those which use it, normally think of user cost as being positive. Moreover, it is difficult to conceive of a case where *marginal* user cost associated with an increase in A, i.e. $\dfrac{dU}{dA}$, will be other than positive.

It may be convenient to mention here, in anticipation of the latter part of this chapter, that, for the community as a whole, the aggregate *consumption* (C) of the period is equal to $\Sigma(A - A_1)$, and the aggregate *investment* (I) is equal to $\Sigma(A_1 - U)$. Moreover, U is the individual entrepreneur's disinvestment (and $-U$ his investment) in respect of his own equipment exclusive of what he buys from other entrepreneurs. Thus in a completely integrated system (where $A_1 = 0$) consumption is equal to A and investment to $-U$, i.e. to $G - (G' - B')$. The slight complication of the above, through the introduction of A_1, is simply due to the desirability of providing in a generalised way for the case of a non-integrated system of production.

Furthermore, the *effective demand* is simply the aggregate income (or proceeds) which the entrepreneurs expect to receive, inclusive of the incomes which they will hand on to the other factors of production, from the amount of current employment which they decide to give. The aggregate demand function relates various hypothetical quantities of employment to the proceeds which their outputs are expected to yield; and the effective demand is the point on the aggregate demand function which becomes effective because, taken in conjunction with the conditions of supply, it corresponds to the level of employment which maximises the entrepreneur's expectation of profit.

This set of definitions also has the advantage that we can equate the marginal proceeds (or income) to the marginal factor cost; and thus arrive at the same sort of propositions relating marginal proceeds thus defined to marginal factor costs as have been stated by those economists who, by ignoring user cost or assuming it to be zero, have equated supply price[1] to marginal factor cost.[2]

---

[1] *Supply price* is, I think, an incompletely defined term, if the problem of defining user cost has been ignored. The matter is further discussed in the appendix to this chapter, where I argue that the exclusion of user cost from supply price, whilst sometimes appropriate in the case of aggregate supply price, is inappropriate to the problems of the supply price of a unit of output for an individual firm.

[2] For example, let us take $Z_w = \phi(N)$, or alternatively $Z = W . \phi(N)$ as the aggregate supply function (where W is the wage-unit and $W . Z_w = Z$). Then, since the proceeds of the marginal product is equal to the marginal factor-cost at every point on the aggregate supply curve, we have

$$\Delta N = \Delta A_w - \Delta U_w = \Delta Z_w = \Delta \phi(N),$$

that is to say $\phi'(N) = 1$; provided that factor cost bears a constant ratio to wage cost, and that the aggregate supply function for each firm (the number

(ii) We turn, next, to the second of the principles referred to above. We have dealt so far with that part of the change in the value of the capital equipment at the end of the period as compared with its value at the beginning which is due to the *voluntary* decisions of the entrepreneur in seeking to maximise his profit. But there may, in addition, be an *involuntary* loss (or gain) in the value of his capital equipment, occurring for reasons beyond his control and irrespective of his current decisions, on account of (e.g.) a change in market values, wastage by obsolescence or the mere passage of time, or destruction by catastrophe such as war or earthquake. Now some part of these involuntary losses, whilst they are unavoidable, are—broadly speaking—not unexpected; such as losses through the lapse of time irrespective of use, and also 'normal' obsolescence which, as Professor Pigou expresses it, 'is sufficiently regular to be foreseen, if not in detail, at least in the large', including, we may add, those losses to the community as a whole which are sufficiently regular to be commonly regarded as 'insurable risks'. Let us ignore for the moment the fact that the amount of the expected loss depends on when the expectation is assumed to be framed, and let us call the depreciation of the equipment, which is involuntary but not unexpected, i.e. the excess of the expected depreciation over the user cost, the *supplementary cost*, which will be written V. It is, perhaps, hardly necessary to point out that this definition is not the same as Marshall's definition of supplementary cost, though the underlying idea, namely, of dealing with that part of the expected depreciation which does not enter into prime cost, is similar.

of which is assumed to be constant) is independent of the number of men employed in other industries, so that the terms of the above equation, which hold good for each individual entrepreneur, can be summed for the entrepreneurs as a whole. This means that, if wages are constant and other factor costs are a constant proportion of the wages-bill, the aggregate supply function is linear with a slope given by the reciprocal of the money-wage.

In reckoning, therefore, the *net income* and the *net profit* of the entrepreneur it is usual to deduct the estimated amount of the supplementary cost from his income and gross profit as defined above. For the psychological effect on the entrepreneur, when he is considering what he is free to spend and to save, of the supplementary cost is virtually the same as though it came off his gross profit. In his capacity as a *producer* deciding whether or not to use the equipment, prime cost and gross profit, as defined above, are the significant concepts. But in his capacity as a *consumer* the amount of the supplementary cost works on his mind in the same way as if it were a part of the prime cost. Hence we shall not only come nearest to common usage but will also arrive at a concept which is relevant to the amount of consumption, if, in defining aggregate *net* income, we deduct the supplementary cost as well as the user cost, so that aggregate *net income* is equal to $A - U - V$.

There remains the change in the value of the equipment, due to unforeseen changes in market values, exceptional obsolescence or destruction by catastrophe, which is both involuntary and—in a broad sense—unforeseen. The actual loss under this head, which we disregard even in reckoning net income and charge to capital account, may be called the *windfall loss*.

The *causal* significance of net income lies in the psychological influence of the magnitude of V on the amount of current consumption, since *net income* is what we suppose the ordinary man to reckon his available income to be when he is deciding how much to spend on current consumption. This is not, of course, the only factor of which he takes account when he is deciding how much to spend. It makes a considerable difference, for example, how much windfall gain or loss he is making on capital account. But there is a difference between the supplementary cost and a windfall loss in that changes in the former are apt to affect him *in just the same way* as changes in his gross profit. It is the excess of the proceeds of the current output over the *sum* of the prime cost and the supplementary cost which is relevant to the entrepreneur's consumption; whereas, although the windfall loss (or gain) enters into his decisions, it does not enter into them on the same scale—a given windfall loss does not have the same effect as an equal supplementary cost.

We must now recur, however, to the point that the line between supplementary costs and windfall losses, i.e. between those unavoidable losses which we think it proper to debit to income account and those which it is reasonable to reckon as a windfall loss (or gain) on capital account, is partly a conventional or psychological one, depending on what are the commonly accepted criteria for estimating the former. For no unique principle can be established for the estimation of supplementary cost, and its amount will depend on our choice of an accounting method. The expected value of the supplementary cost, when the equipment was originally produced, is a definite quantity. But if it is re-estimated subsequently, its amount over the remainder of the life of the equipment may have changed as a result of a change in the meantime in our expectations; the windfall capital loss being the discounted value of the difference between the former and the revised expectation of the prospective series of $U + V$. It is a widely approved principle of business accounting, endorsed by the Inland Revenue authorities, to establish a figure for the sum of the supplementary cost and the user cost when the equipment is acquired and to maintain this unaltered during the life of the equipment, irrespective of subsequent changes in expectation. In this case the supplementary cost over any period must be taken as the excess of this predetermined figure over the actual user cost. This has the advantage of ensuring that the windfall gain or loss shall be zero over the life of the equipment taken as a whole. But it is also reasonable in certain circumstances to recalculate the allowance for supplementary cost on the basis of current values and expectations at an arbitrary accounting interval, e.g. annually. Business men in fact differ as to which course they adopt. It may be convenient to call the initial expectation of supplementary cost when the equipment is first acquired the *basic supplementary cost*, and the same quantity recalculated up to date on the basis of current values and expectations the *current supplementary cost.*

Thus we cannot get closer to a quantitative definition of supplementary cost than that it comprises those deductions from his income which a typical entrepreneur makes before reckoning what he considers his *net* income for the purpose of declaring a dividend (in the case of a corporation) or of deciding the scale of his current consumption (in the case of an individual). Since windfall charges on capital account are not going to be ruled out of the picture, it is clearly better, in case of doubt, to assign an item to capital account, and to include in supplementary cost only what rather obviously belongs there. For any overloading of the former can be corrected by allowing it more influence on the rate of current consumption than it would otherwise have had.

It will be seen that our definition of *net income* comes very close to Marshall's definition of *income*, when he decided to take refuge in the practices of the Income Tax Commissioners and—broadly speaking to regard as income whatever they, with their experience, choose to treat as such. For the fabric of their decisions can be regarded as the result of the most careful and extensive investigation which is available, to interpret what, in practice, it is usual to treat as net income. It also corresponds to the money value of Professor Pigou's most recent definition of the national dividend.[1]

---

[1] *Economic Journal,* June 1935, p. 235.

It remains true, however, that net income, being based on an equivocal criterion which different authorities might interpret differently, is not perfectly clear-cut. Professor Hayek, for example, has suggested that an individual owner of capital goods might aim at keeping the income he derives from his possessions constant, so that he would not feel himself free to spend his income on consumption until he had set aside sufficient to offset any tendency of his investment-income to decline for whatever reason.[1] I doubt if such an individual exists; but, obviously, no theoretical objection can be raised against this deduction as providing a possible psychological criterion of net income. But when Professor Hayek infers that the concepts of saving and investment suffer from a corresponding vagueness, he is only right if he means *net saving* and *net investment*. The *saving* and the *investment*, which are relevant to the theory of employment, are clear of this defect, and are capable of objective definition, as we have shown above.

Thus it is a mistake to put all the emphasis on *net income*, which is only relevant to decisions concerning consumption, and is, moreover, only separated from various other factors affecting consumption by a narrow line; and to overlook (as has been usual) the concept of *income* proper, which is the concept relevant to decisions concerning current production and is quite unambiguous.

The above definitions of income and of net income are intended to conform as closely as possible to common usage. It is necessary, therefore, that I should at once remind the reader that in my *Treatise on Money* I defined income in a special sense. The peculiarity in my former definition related to that part of aggregate income which accrues to the entrepreneurs, since I took neither the profit (whether gross or net) actually realised from their current operations nor the profit which they expected when they decided to under-

[1] 'The Maintenance of Capital', *Economica*, August 1935, p. 241 *et seq.*

take their current operations, but in some sense (not, as I now think, sufficiently defined if we allow for the possibility of changes in the scale of output) a normal or equilibrium profit; with the result that on this definition saving exceeded investment by the amount of the excess of normal profit over the actual profit. I am afraid that this use of terms has caused considerable confusion, especially in the case of the correlative use of saving; since conclusions (relating, in particular, to the excess of saving over investment), which were only valid if the terms employed were interpreted in my special sense, have been frequently adopted in popular discussion as though the terms were being employed in their more familiar sense. For this reason, and also because I no longer require my former terms to express my ideas accurately, I have decided to discard them— with much regret for the confusion which they have caused.

II. *Saving and Investment*

Amidst the welter of divergent usages of terms, it is agreeable to discover one fixed point. So far as I know, everyone is agreed that *saving* means the excess of income over expenditure on consumption. Thus any doubts about the meaning of *saving* must arise from doubts about the meaning either of *income* or of *consumption*. *Income* we have defined above. Expenditure on consumption during any period must mean the value of goods sold to consumers during that period, which throws us back to the question of what is meant by a consumer-purchaser. Any reasonable definition of the line between consumer-purchasers and investor-purchasers will serve us equally well, provided that it is consistently applied. Such problem as there is, e.g. whether it is right to regard the purchase of a motor-car as a consumer-purchase and the purchase of a house as an investor-purchase, has been frequently discussed and I have nothing material to add to the discussion.

The criterion must obviously correspond to where we draw the line between the consumer and the entrepreneur. Thus when we have defined A_1 as the value of what one entrepreneur has purchased from another, we have implicitly settled the question. It follows that expenditure on consumption can be unambiguously defined as $\Sigma(A - A_1)$, where ΣA is the total sales made during the period and ΣA_1 is the total sales made by one entrepreneur to another. In what follows it will be convenient, as a rule, to omit Σ and write A for the aggregate sales of all kinds, A_1 for the aggregate sales from one entrepreneur to another and U for the aggregate user costs of the entrepreneurs.

Having now defined both *income* and *consumption*, the definition of *saving*, which is the excess of income over consumption, naturally follows. Since income is equal to $A - U$ and consumption is equal to $A - A_1$, it follows that saving is equal to $A_1 - U$. Similarly, we have *net saving* for the excess of *net income* over consumption, equal to $A_1 - U - V$.

Our definition of income also leads at once to the definition of *current investment*. For we must mean by this the current addition to the value of the capital equipment which has resulted from the productive activity of the period. This is, clearly, equal to what we have just defined as saving. For it is that part of the income of the period which has not passed into consumption. We have seen above that as the result of the production of any period entrepreneurs end up with having sold finished output having a value A and with a capital equipment which has suffered a deterioration measured by U (or an improvement measured by $-U$ where U is negative) as a result of having produced and parted with A, after allowing for purchases A_1 from other entrepreneurs. During the same period finished output having a value $A - A_1$ will have passed into consumption. The excess of $A - U$ over $A - A_1$, namely $A_1 - U$, is the addition to capital equipment as a result of the productive activities of the period and is, therefore, the *investment* of the period. Similarly $A_1 - U - V$, which is the *net* addition to capital equipment, after allowing for normal impairment in the value of capital apart from its being used and apart from windfall changes in the value of the equipment chargeable to capital account, is the *net investment* of the period.

Whilst, therefore, the amount of saving is an outcome of the collective behaviour of individual consumers and the amount of investment of the collective behaviour of individual entrepreneurs, these two amounts are necessarily equal, since each of them is equal to the excess of income over consumption. Moreover, this conclusion in no way depends on any subtleties or peculiarities in the definition of income given above. Provided it is agreed that income is equal to the value of current output, that current investment is equal to the value of that part of current output which is not consumed, and that saving is equal to the excess of income over consumption—all of which is conformable both to common sense and to the traditional usage of the great majority of economists—the equality of saving and investment necessarily follows. In short—

Income = value of output = consumption + investment.
Saving = income − consumption.
Therefore saving = investment.

Thus *any* set of definitions which satisfy the above conditions leads to the same conclusion. It is only by denying the validity of one or other of them that the conclusion can be avoided.

The equivalence between the quantity of saving and the quantity of investment emerges from the *bilateral* character of the transactions between the producer on the one hand and, on the other hand, the consumer or the purchaser of capital equipment.

Income is created by the value in excess of user cost which the producer obtains for the output he has sold; but the whole of this output must obviously have been sold either to a consumer or to another entrepreneur; and each entrepreneur's current investment is equal to the excess of the equipment which he has purchased from other entrepreneurs over his own user cost. Hence, in the aggregate the excess of income over consumption, which we call saving, cannot differ from the addition to capital equipment which we call investment. And similarly with net saving and net investment. Saving, in fact, is a mere residual. The decisions to consume and the decisions to invest between them determine incomes. Assuming that the decisions to invest become effective, they must in doing so either curtail consumption or expand income. Thus the act of investment in itself cannot help causing the residual or margin, which we call saving, to increase by a corresponding amount.

It might be, of course, that individuals were so *tête montée* in their decisions as to how much they themselves would save and invest respectively, that there would be no point of price equilibrium at which transactions could take place. In this case our terms would cease to be applicable, since output would no longer have a definite market value, prices would find no resting-place between zero and infinity. Experience shows, however, that this, in fact, is not so; and that there are habits of psychological response which allow of an equilibrium being reached at which the readiness to buy is equal to the readiness to sell. That there should be such a thing as a market value for output is, at the same time, a necessary condition for money-income to possess a definite value and a sufficient condition for the aggregate amount which saving individuals decide to save to be equal to the aggregate amount which investing individuals decide to invest.

Clearness of mind on this matter is best reached,

perhaps, by thinking in terms of decisions to consume (or to refrain from consuming) rather than of decisions to save. A decision to consume or not to consume truly lies within the power of the individual; so does a decision to invest or not to invest. The amounts of aggregate income and of aggregate saving are the *results* of the free choices of individuals whether or not to consume and whether or not to invest; but they are neither of them capable of assuming an independent value resulting from a separate set of decisions taken irrespective of the decisions concerning consumption and investment. In accordance with this principle, the conception of the *propensity to consume* will, in what follows, take the place of the propensity or disposition to save.

Appendix on User Cost

User cost has, I think, an importance for the classical theory of value which has been overlooked. There is more to be said about it than would be relevant or appropriate in this place. But, as a digression, we will examine it somewhat further in this appendix.

An entrepreneur's user cost is by definition equal to

$$A_1 + (G' - B') - G,$$

where A_1 is the amount of our entrepreneur's purchases from other entrepreneurs, G the actual value of his capital equipment at the end of the period, and G' the value it might have had at the end of the period if he had refrained from using it and had spent the optimum sum B' on its maintenance and improvement. Now $G - (G' - B')$, namely the increment in the value of the entrepreneur's equipment beyond the net value which he has inherited from the previous period, represents the entrepreneur's current investment in his equipment and can be written I. Thus U, the user cost of his sales-turnover A, is equal to $A_1 - I$ where A_1 is what he has bought from other entrepreneurs and I is what he has currently invested in his own equipment. A little reflection will show that all this is no more than common sense. Some part of his outgoings to other entrepreneurs is balanced by the value of his current investment in his own equipment, and the rest represents the sacrifice which the output he has sold must have cost him over and above the total sum which he has paid out to the factors of production. If the reader tries to express the substance of this otherwise, he will find that its advantage lies in its avoidance of insoluble (and unnecessary) accounting problems. There is, I think, no other way of analysing the current proceeds of production unambiguously. If industry is completely integrated or if the entrepreneur has bought nothing from outside, so that $A_1 = 0$, the user cost is simply the equivalent of the current disinvestment involved in using the equipment; but we are still left with the advantage that we do not require at any stage of the analysis to allocate the factor cost between the goods which are sold and the equipment which is retained. Thus we can regard the employment given by a firm, whether integrated or individual, as depending on a single consolidated decision—a procedure which corresponds to the actual interlocking character of the production of what is currently sold with total production.

The concept of user cost enables us, moreover, to give a clearer definition than that usually adopted of the short-period supply price of a unit of a firm's saleable output. For the short-period supply price is the sum of the marginal factor cost and the marginal user cost.

Now in the modern theory of value it has been a usual practice to equate the short-period supply price to the marginal factor cost alone. It is obvious, however, that this is only legitimate if marginal user cost is zero or if supply price is specially defined so as to be net of marginal user cost, just as I have defined (p. 24 above) 'proceeds' and 'aggregate supply price' as being net of aggregate user cost. But, whereas it may be occasionally convenient in dealing with *output as a whole* to deduct user cost, this procedure deprives our analysis of all reality if it is habitually (and tacitly) applied to the output of a single industry or firm, since it divorces the 'supply price' of an article from any ordinary sense of its 'price'; and some confusion may have resulted from the practice of doing so. It seems to have been assumed that 'supply price' has an obvious meaning as applied to a unit of the saleable output of an individual firm, and the matter has not been deemed to require discussion. Yet the treatment both of what is purchased from other firms and of the wastage of the firm's own equipment as a consequence of producing the marginal output involves the whole pack of perplexities which attend the definition of income. For, even if we assume that the marginal cost of purchases from other firms involved in selling an additional unit of output has to be deducted from the sale-proceeds per unit in order to give us what we mean by our firm's supply price, we still have to allow for the marginal disinvestment in the firm's own equipment involved in producing the marginal output. Even if all production is carried on by a completely integrated firm, it is still illegitimate to suppose that the marginal user cost is zero, i.e. that the marginal disinvestment in equipment due to the production of the marginal output can generally be neglected.

The concepts of user cost and of supplementary cost also enable us to establish a clearer relationship between long-period supply price and short-period supply price. Long-period cost must obviously include an amount to cover the basic supplementary cost as well as the expected prime cost appropriately averaged over the life of the equipment. That is to say, the long-period cost of the output is equal to the expected sum of the prime cost and the supplementary cost; and, furthermore, in order to yield a normal profit, the long-period supply price must exceed the long-period cost thus calculated by an amount determined by the current rate of interest on loans of comparable term and risk, reckoned as a percentage of the cost of the equipment. Or if we prefer to take a standard 'pure' rate of interest, we must include in the long-period cost a third term which we might call the *risk-cost* to cover the unknown possibilities of the actual yield differing from the expected yield. Thus the long-period supply price is equal to the sum of the prime cost, the supplementary cost, the risk cost and the interest cost, into which several components it can be analysed. The short-period supply price, on the other hand, is equal to the *marginal* prime cost. The entrepreneur must, therefore, expect, when he buys or constructs his equipment, to cover his supplementary cost, his risk cost and his interest cost out of the excess of the marginal value of the prime cost over its average value; so that in long-period equilibrium the excess of the marginal prime cost over the average prime cost is equal to the sum of the supplementary, risk and interest costs.[1]

The level of output, at which marginal prime cost is exactly equal to the sum of the average prime and supplementary costs, has a special importance, because it is the point at which the entrepreneur's trading account breaks even. It corresponds,

---

[1] This way of putting it depends on the convenient assumption that the marginal prime cost curve is continuous throughout its length for changes in output. In fact, this assumption is often unrealistic, and there may be one or more points of discontinuity, especially when we reach an output corresponding to the technical full capacity of the equipment. In this case the marginal analysis partially breaks down; and the price may *exceed* the marginal prime cost, where the latter is reckoned in respect of a small *decrease* of output. (Similarly there may often be a discontinuity in the downward direction, i.e. for a reduction in output *below* a certain point.) This is important when we are considering the short-period supply price in long-period equilibrium, since in that case any discontinuities, which may exist corresponding to a point of technical full capacity, must be supposed to be in operation. Thus the short-period supply price in long-period equilibrium may have to exceed the marginal prime cost (reckoned in terms of a small *decrease* of output).

that is to say, to the point of zero net profit; whilst with a smaller output than this he is trading at a net loss.

The extent to which the supplementary cost has to be provided for apart from the prime cost varies very much from one type of equipment to another. Two extreme cases are the following:

(i) Some part of the maintenance of the equipment must necessarily take place *pari passu* with the act of using it (e.g. oiling the machine). The expense of this (apart from outside purchases) is included in the factor cost. If, for physical reasons, the exact amount of the whole of the current depreciation has necessarily to be made good in this way, the amount of the user cost (apart from outside purchases) would be equal and opposite to that of the supplementary cost; and in long-period equilibrium the marginal factor cost would exceed the average factor cost by an amount equal to the risk and interest cost.

(ii) Some part of the deterioration in the value of the equipment only occurs if it is used. The cost of this is charged in user cost, in so far as it is not made good *pari passu* with the act of using it. If loss in the value of the equipment could only occur in this way, supplementary cost would be zero.

It may be worth pointing out that an entrepreneur does not use his oldest and worst equipment first, merely because its user cost is low; since its low user cost may be outweighed by its relative inefficiency, i.e. by its high factor cost. Thus an entrepreneur uses by preference that part of his equipment for which the user cost *plus* factor cost is least per unit of output.[1] It follows that for any given volume of output of the product in question there is a corresponding user cost,[2] but that this total user cost does not bear a uniform relation to the marginal user cost, i.e. to the increment of user cost due to an increment in the rate of output.

II

User cost constitutes one of the links between the present and the future. For in deciding his scale of production an

---

[1] Since user cost partly depends on expectations as to the future level of wages, a reduction in the wage–unit which is expected to be short-lived will cause factor cost and user cost to move in different proportions and so affect what equipment is used, and, conceivably, the level of effective demand, since factor cost may enter into the determination of effective demand in a different way from user cost.

[2] The user cost of the equipment which is first brought into use is not necessarily independent of the total volume of output (see below); i.e. the user cost may be affected all along the line when the total volume of output is changed.

entrepreneur has to exercise a choice between using up his equipment now and preserving it to be used later on. It is the expected sacrifice of future benefit involved in present use which determines the amount of the user cost, and it is the marginal amount of this sacrifice which, together with the marginal factor cost and the expectation of the marginal proceeds, determines his scale of production. How, then, is the user cost of an act of production calculated by the entrepreneur?

We have defined the user cost as the reduction in the value of the equipment due to using it as compared with not using it, after allowing for the cost of the maintenance and improvements which it would be worth while to undertake and for purchases from other entrepreneurs. It must be arrived at, therefore, by calculating the discounted value of the additional prospective yield which would be obtained at some later date if it were not used now. Now this must be at least equal to the present value of the opportunity to postpone replacement which will result from laying up the equipment; and it may be more.[1]

If there is no surplus or redundant stock, so that more units of similar equipment are being newly produced every year either as an addition or in replacement, it is evident that marginal user cost will be calculable by reference to the amount by which the life or efficiency of the equipment will be shortened if it is used, and the current replacement cost. If, however, there is redundant equipment, then the user cost will also depend on the rate of interest and the current (i.e. re-estimated) supplementary cost over the period of time before the redundancy is expected to be absorbed through wastage, etc. In this way interest cost and current supplementary cost enter indirectly into the calculation of user cost.

The calculation is exhibited in its simplest and most intelligible form when the factor cost is zero, e.g. in the case of a redundant stock of a raw material such as copper, on the lines which I have worked out in my *Treatise on Money*, vol. ii. chap. 29. Let us take the prospective values of copper at various future dates, a series which will be governed by the rate at which redundancy is being absorbed and gradually approaches the estimated normal cost. The present value or user cost of a ton of surplus copper will then be equal to the greatest of the values obtainable by subtracting from the estimated future value at any

---

[1] It will be more when it is expected that a more than normal yield can be obtained at some later date, which, however, is not expected to last long enough to justify (or give time for) the production of new equipment. To-day's user cost is equal to the maximum of the discounted values of the potential expected yields of all the to-morrows.

given date of a ton of copper the interest cost and the current supplementary cost on a ton of copper between that date and the present.

In the same way the user cost of a ship or factory or machine, when these equipments are in redundant supply, is its estimated replacement cost discounted at the percentage rate of its interest and current supplementary costs to the prospective date of absorption of the redundancy.

We have assumed above that the equipment will be replaced in due course by an identical article. If the equipment in question will not be renewed identically when it is worn out, then its user cost has to be calculated by taking a proportion of the user cost of the new equipment, which will be erected to do its work when it is discarded, given by its comparative efficiency.

III

The reader should notice that, where the equipment is not obsolescent but merely redundant for the time being, the difference between the actual user cost and its normal value (i.e. the value when there is no redundant equipment) varies with the interval of time which is expected to elapse before the redundancy is absorbed. Thus if the type of equipment in question is of all ages and not 'bunched', so that a fair proportion is reaching the end of its life annually, the marginal user cost will not decline greatly unless the redundancy is exceptionally excessive. In the case of a general slump, marginal user cost will depend on how long entrepreneurs expect the slump to last. Thus the rise in the supply price when affairs begin to mend may be partly due to a sharp increase in marginal user cost due to a revision of their expectations.

It has sometimes been argued, contrary to the opinion of business men, that organised schemes for scrapping redundant plant cannot have the desired effect of raising prices unless they apply to the *whole* of the redundant plant. But the concept of user cost shows how the scrapping of (say) half the redundant plant may have the effect of raising prices immediately. For by bringing the date of the absorption of the redundancy nearer, this policy raises marginal user cost and consequently increases the current supply price. Thus business men would seem to have the notion of user cost implicitly in mind, though they do not formulate it distinctly.

If the supplementary cost is heavy, it follows that the marginal user cost will be low when there is surplus equipment. Moreover, when there is surplus equipment, the marginal factor and user costs are unlikely to be much in excess of their average value. If both these conditions are fulfilled, the existence of surplus equipment is likely to lead to the entrepreneur's working at a net loss, and perhaps at a heavy net loss. There will not be a sudden transition from this state of affairs to a normal profit, taking place at the moment when the redundancy is absorbed. As the redundancy becomes less, the user cost will gradually increase; and the excess of marginal over average factor and user cost may also gradually increase.

IV

In Marshall's *Principles of Economics* (6th ed. p. 360) a part of user cost is included in prime cost under the heading of 'extra wear-and-tear of plant'. But no guidance is given as to how this item is to be calculated or as to its importance. In his *Theory of Unemployment* (p. 42) Professor Pigou expressly assumes that the marginal disinvestment in equipment due to the marginal output can, in general, be neglected: 'The differences in the quantity of wear-and-tear suffered by equipment and in the costs of non-manual labour employed, that are associated with differences in output, are ignored, as being, in general, of secondary importance'.[1] Indeed, the notion that the disinvestment in equipment is zero at the margin of production runs through a good deal of recent economic theory. But the whole problem is brought to an obvious head as soon as it is thought necessary to explain exactly what is meant by the supply price of an individual firm.

It is true that the cost of maintenance of idle plant may often, for the reasons given above, reduce the magnitude of marginal user cost, especially in a slump which is expected to last a long time. Nevertheless a very low user cost at the margin is not a characteristic of the short period as such, but of particular situations and types of equipment where the cost of maintaining idle plant happens to be heavy, and of those disequilibria which are characterised by very rapid obsolescence or great redundancy, especially if it is coupled with a large proportion of comparatively new plant.

In the case of raw materials the necessity of allowing for user cost is obvious;—if a ton of copper is used up to-day it cannot be

[1] Mr Hawtrey (*Economica*, May 1934, p. 145) has called attention to Prof. Pigou's identification of supply price with marginal labour cost, and has contended that Prof. Pigou's argument is thereby seriously vitiated.

72

used to-morrow, and the value which the copper would have for the purposes of to-morrow must clearly be reckoned as a part of the marginal cost. But the fact has been overlooked that copper is only an extreme case of what occurs whenever capital equipment is used to produce. The assumption that there is a sharp division between raw materials where we must allow for the disinvestment due to using them and fixed capital where we can safely neglect it does not correspond to the facts;—especially in normal conditions where equipment is falling due for replacement every year and the use of equipment brings nearer the date at which replacement is necessary.

It is an advantage of the concepts of user cost and supplementary cost that they are as applicable to working and liquid capital as to fixed capital. The essential difference between raw materials and fixed capital lies not in their liability to user and supplementary costs, but in the fact that the return to liquid capital consists of a single term; whereas in the case of fixed capital, which is durable and used up gradually, the return consists of a series of user costs and profits earned in successive periods.

73

Chapter 7

THE MEANING OF
SAVING AND INVESTMENT
FURTHER CONSIDERED

I

In the previous chapter, *saving* and *investment* have been so defined that they are necessarily equal in amount, being, for the community as a whole, merely different aspects of the same thing. Several contemporary writers (including myself in my *Treatise on Money*) have, however, given special definitions of these terms on which they are not necessarily equal. Others have written on the assumption that they may be unequal without prefacing their discussion with any definitions at all. It will be useful, therefore, with a view to relating the foregoing to other discussions of these terms, to classify some of the various uses of them which appear to be current.

So far as I know, everyone agrees in meaning by *saving* the excess of income over what is spent on consumption. It would certainly be very inconvenient and misleading not to mean this. Nor is there any important difference of opinion as to what is meant by expenditure on consumption. Thus the differences of usage arise either out of the definition of *investment* or out of that of *income*.

II

Let us take *investment* first. In popular usage it is common to mean by this the purchase of an asset, old or new, by an individual or a corporation. Occasionally, the term might be restricted to the purchase of an asset on the Stock Exchange. But we speak just as readily of investing, for example, in a house, or in a machine, or in a stock of finished or unfinished goods; and, broadly speaking, new investment, as distinguished from reinvestment, means the purchase of a capital asset of any kind out of income. If we reckon the sale of an investment as being negative investment, i.e. disinvestment, my own definition is in accordance with popular usage; since exchanges of old investments necessarily cancel out. We have, indeed, to adjust for the creation and discharge of debts (including changes in the quantity of credit or money); but since for the community as a whole the increase or decrease of the aggregate creditor position is always exactly equal to the increase or decrease of the aggregate debtor position, this complication also cancels out when we are dealing with aggregate investment. Thus, assuming that income in the popular sense corresponds to my net income, aggregate investment in the popular sense coincides with my definition of net investment, namely the net addition to all kinds of capital equipment, after allowing for those changes in the value of the old capital equipment which are taken into account in reckoning net income.

Investment, thus defined, includes, therefore, the increment of capital equipment, whether it consists of fixed capital, working capital or liquid capital; and the significant differences of definition (apart from the distinction between investment and net investment) are due to the exclusion from investment of one or more of these categories.

Mr Hawtrey, for example, who attaches great importance to changes in liquid capital, i.e. to undesigned increments (or decrements) in the stock of unsold goods, has suggested a possible definition of investment from which such changes are excluded. In this case an excess of saving over investment would be the same thing as an undesigned increment in the stock of unsold goods, i.e. as an increase of liquid capital. Mr Hawtrey has not convinced me that this is the factor to stress; for it lays all the emphasis on the correction of changes which were in the first instance unforeseen, as compared with those which are, rightly or wrongly, anticipated. Mr Hawtrey regards the daily decisions of entrepreneurs concerning their scale of output as being varied from the scale of the previous day by reference to the changes in their stock of unsold goods. Certainly, in the case of consumption goods, this plays an important part in their decisions. But I see no object in excluding the play of other factors on their decisions; and I prefer, therefore, to emphasise the total change of effective demand and not merely that part of the change in effective demand which reflects the increase or decrease of unsold stocks in the previous period. Moreover, in the case of fixed capital, the increase or decrease of unused capacity corresponds to the increase or decrease in unsold stocks in its effect on decisions to produce; and I do not see how Mr Hawtrey's method can handle this at least equally important factor.

It seems probable that capital formation and capital consumption, as used by the Austrian school of economists, are not identical either with investment and disinvestment as defined above or with net investment and disinvestment. In particular, capital consumption is said to occur in circumstances where there is quite clearly no net decrease in capital equipment as defined above. I have, however, been unable to discover a reference to any passage where the meaning of these terms is clearly explained. The statement, for example, that capital formation occurs when there is a lengthening of the period of production does not much advance matters.

76

III

We come next to the divergences between saving and investment which are due to a special definition of income and hence of the excess of income over consumption. My own use of terms in my *Treatise on Money* is an example of this. For, as I have explained on p. 60 above, the definition of income, which I there employed, differed from my present definition by reckoning as the income of entrepreneurs not their actually realised profits but (in some sense) their 'normal profit'. Thus by an excess of saving over investment I meant that the scale of output was such that entrepreneurs were earning a less than normal profit from their ownership of the capital equipment; and by an increased excess of saving over investment I meant that a decline was taking place in the actual profits, so that they would be under a motive to contract output.

As I now think, the volume of employment (and consequently of output and real income) is fixed by the entrepreneur under the motive of seeking to maximise his present and prospective profits (the allowance for user cost being determined by his view as to the use of equipment which will maximise his return from it over its whole life); whilst the volume of employment which will maximise his profit depends on the aggregate demand function given by his expectations of the sum of the proceeds resulting from consumption and investment respectively on various hypotheses. In my *Treatise on Money* the concept of *changes* in the excess of investment over saving, as there defined, was a way of handling changes in profit, though I did not in that book distinguish clearly between expected and realised results.[1] I there argued that change in the excess of

---

[1] My method there was to regard the current realised profit as determining the current expectation of profit.

investment over saving was the motive force governing changes in the volume of output. Thus the new argument, though (as I now think) much more accurate and instructive, is essentially a development of the old. Expressed in the language of my *Treatise on Money*, it would run: the expectation of an increased excess of investment over saving, given the former volume of employment and output, will induce entrepreneurs to increase the volume of employment and output. The significance of both my present and my former arguments lies in their attempt to show that the volume of employment is determined by the estimates of effective demand made by the entrepreneurs, an expected increase of investment relatively to saving as defined in my *Treatise on Money* being a criterion of an increase in effective demand. But the exposition in my *Treatise on Money* is, of course, very confusing and incomplete in the light of the further developments here set forth.

Mr D. H. Robertson has defined to-day's income as being equal to *yesterday's* consumption *plus* investment, so that to-day's saving, in his sense, is equal to yesterday's investment *plus* the excess of yesterday's consumption over to-day's consumption. On this definition saving can exceed investment, namely, by the excess of yesterday's income (in my sense) over to-day's income. Thus when Mr Robertson says that there is an excess of saving over investment, he means literally the same thing as I mean when I say that income is falling, and the excess of saving in his sense is exactly equal to the decline of income in my sense. If it were true that current expectations were always determined by yesterday's realised results, to-day's effective demand would be equal to yesterday's income. Thus Mr Robertson's method might be regarded as an alternative· attempt to mine (being, perhaps, a first approximation to it) to make the same distinction, so vital for causal analysis, that I have tried to make by the contrast between effective demand and income.[1]

IV

We come next to the much vaguer ideas associated with the phrase 'forced saving'. Is any clear significance discoverable in these? In my *Treatise on Money* (vol. I, p. 171, footnote [JMK, vol. v, p. 154] I gave some references to earlier uses of this phrase and suggested that they bore some affinity to the difference between investment and 'saving' in the sense in which I there used the latter term. I am no longer confident that there was in fact so much affinity as I then supposed. In any case, I feel sure that 'forced saving' and analogous phrases employed more recently (e.g. by Professor Hayek or Professor Robbins) have no definite relation to the difference between investment and 'saving' in the sense intended in my *Treatise on Money*. For whilst these authors have not explained exactly what they mean by this term, it is clear that 'forced saving', in their sense, is a phenomenon which results directly from, and is measured by, changes in the quantity of money or bank-credit.

It is evident that a change in the volume of output and employment will, indeed, cause a change in income measured in wage-units; that a change in the wage-unit will cause both a redistribution of income between borrowers and lenders and a change in aggregate income measured in money; and that in either event there will (or may) be a change in the amount saved. Since, therefore, changes in the quantity of money may result, through their effect on the rate of interest, in a change in the volume and distribution of income (as we shall show later), such changes may involve, indirectly, a change in the amount saved. But such

[1] *Vide* Mr Robertson's article 'Saving and Hoarding' (*Economic Journal*, September 1933, p. 399) and the discussion between Mr Robertson, Mr Hawtrey and myself (*Economic Journal*, December 1933, p. 658) [*JMK*, vol. XIII].

changes in the amounts saved are no more 'forced savings' than any other changes in the amounts saved due to a change in circumstances; and there is no means of distinguishing between one case and another, unless we specify the amount saved in certain given conditions as our norm or standard. Moreover, as we shall see, the amount of the change in aggregate saving which results from a given change in the quantity of money is highly variable and depends on many other factors.

Thus 'forced saving' has no meaning until we have specified some standard rate of saving. If we select (as might be reasonable) the rate of saving which corresponds to an established state of full employment, the above definition would become: 'Forced saving is the excess of actual saving over what would be saved if there were full employment in a position of long-period equilibrium'. This definition would make good sense, but a sense in which a forced excess of saving would be a very rare and a very unstable phenomenon, and a forced *deficiency* of saving the usual state of affairs.

Professor Hayek's interesting 'Note on the Development of the Doctrine of *Forced Saving*'[1] shows that this was in fact the original meaning of the term. 'Forced saving' or 'forced frugality' was, in the first instance, a conception of Bentham's; and Bentham expressly stated that he had in mind the consequences of an increase in the quantity of money (relatively to the quantity of things vendible for money) in circumstances of 'all hands being employed and employed in the most advantageous manner'.[2] In such circumstances, Bentham points out, real income cannot be increased, and, consequently, additional investment, taking place as a result of the transition, involves forced frugality 'at the expense of national comfort and national justice'. All the nineteenth-century

---

[1] *Quarterly Journal of Economics*, November 1932, p. 123.
[2] *Loc. cit.* p. 125.

writers who dealt with this matter had virtually the same idea in mind. But an attempt to extend this perfectly clear notion to conditions of less than full employment involves difficulties. It is true, of course (owing to the fact of diminishing returns to an increase in the employment applied to a given capital equipment), that *any* increase in employment involves some sacrifice of real income to those who were already employed, but an attempt to relate this loss to the increase in investment which may accompany the increase in employment is not likely to be fruitful. At any rate I am not aware of any attempt having been made by the modern writers who are interested in 'forced saving' to extend the idea to conditions where employment is increasing; and they seem, as a rule, to overlook the fact that the extension of the Benthamite concept of forced frugality to conditions of less than full employment requires some explanation or qualification.

V

The prevalence of the idea that saving and investment, taken in their straightforward sense, can differ from one another, is to be explained, I think, by an optical illusion due to regarding an individual depositor's relation to his bank as being a one-sided transaction, instead of seeing it as the two-sided transaction which it actually is. It is supposed that a depositor and his bank can somehow contrive between them to perform an operation by which savings can disappear into the banking system so that they are lost to investment, or, contrariwise, that the banking system can make it possible for investment to occur, to which no saving corresponds. But no one can save without acquiring an asset, whether it be cash or a debt or capital-goods; and no one can acquire an asset which he did not previously possess, unless *either* an asset of equal value is newly produced *or* someone else

81

parts with an asset of that value which he previously had. In the first alternative there is a corresponding new investment: in the second alternative someone else must be dis-saving an equal sum. For his loss of wealth must be due to his consumption exceeding his income, and not to a loss on capital account through a change in the value of a capital-asset, since it is not a case of his suffering a loss of value which his asset formerly had; he is duly receiving the current value of his asset and yet is not retaining this value in wealth of any form, i.e. he must be spending it on current consumption in excess of current income. Moreover, if it is the banking system which parts with an asset, someone must be parting with cash. It follows that the aggregate saving of the first individual and of others taken together must necessarily be equal to the amount of current new investment.

The notion that the creation of credit by the banking system allows investment to take place to which 'no genuine saving' corresponds can only be the result of isolating one of the consequences of the increased bank-credit to the exclusion of the others. If the grant of a bank credit to an entrepreneur additional to the credits already existing allows him to make an addition to current investment which would not have occurred otherwise, incomes will necessarily be increased and at a rate which will normally *exceed* the rate of increased investment. Moreover, except in conditions of full employment, there will be an increase of real income as well as of money-income. The public will exercise 'a free choice' as to the proportion in which they divide their increase of income between saving and spending; and it is impossible that the intention of the entrepreneur who has borrowed in order to increase investment can become effective (except in substitution for investment by other entrepreneurs which would have occurred otherwise) at a faster rate than the public decide to increase their savings. Moreover, the savings which result from this decision are just as genuine as any other savings. No one can be compelled to own the additional money corresponding to the new bank-credit, unless he deliberately prefers to hold more money rather than some other form of wealth. Yet employment, incomes and prices cannot help moving in such a way that in the new situation someone does choose to hold the additional money. It is true that an unexpected increase of investment in a particular direction may cause an irregularity in the rate of aggregate saving and investment which would not have occurred if it had been sufficiently foreseen. It is also true that the grant of the bank-credit will set up three tendencies —(1) for output to increase, (2) for the marginal product to rise in value in terms of the wage-unit (which in conditions of decreasing return must necessarily accompany an increase of output), and (3) for the wage-unit to rise in terms of money (since this is a frequent concomitant of better employment); and these tendencies may affect the distribution of real income between different groups. But these tendencies are characteristic of a state of increasing output as such, and will occur just as much if the increase in output has been initiated otherwise than by an increase in bank-credit. They can only be avoided by avoiding any course of action capable of improving employment. Much of the above, however, is anticipating the result of discussions which have not yet been reached.

Thus the old-fashioned view that saving always involves investment, though incomplete and misleading, is formally sounder than the new-fangled view that there can be saving without investment or investment without 'genuine' saving. The error lies in proceeding to the plausible inference that, when an individual saves, he will increase aggregate investment by an equal amount. It is true, that, when an individual saves he increases his own wealth. But the conclusion that he also increases aggregate wealth fails to allow for the possibility that an act of individual saving may react on someone else's savings and hence on someone else's wealth.

The reconciliation of the identity between saving and investment with the apparent 'free-will' of the individual to save what he chooses irrespective of what he or others may be investing, essentially depends on saving being, like spending, a two-sided affair. For although the amount of his own saving is unlikely to have any significant influence on his own income, the reactions of the amount of his consumption on the incomes of others makes it impossible for all individuals simultaneously to save any given sums. Every such attempt to save more by reducing consumption will so affect incomes that the attempt necessarily defeats itself. It is, of course, just as impossible for the community as a whole to save *less* than the amount of current investment, since the attempt to do so will necessarily raise incomes to a level at which the sums which individuals choose to save add up to a figure exactly equal to the amount of investment.

The above is closely analogous with the proposition which harmonises the liberty, which every individual possesses, to change, whenever he chooses, the amount of money he holds, with the necessity for the total amount of money, which individual balances add up to, to be exactly equal to the amount of cash which the banking system has created. In this latter case the equality is brought about by the fact that the amount of money which people choose to hold is not independent of their incomes or of the prices of the things (primarily securities), the purchase of which is the natural alternative to holding money. Thus incomes and such prices necessarily change until the aggregate of the amounts of money which individuals choose to hold at the new level of incomes and prices thus brought about has come to equality with the amount of money created by the banking system. This, indeed, is the fundamental proposition of monetary theory.

Both these propositions follow merely from the fact that there cannot be a buyer without a seller or a seller without a buyer. Though an individual whose transactions ·are small in relation to the market can safely neglect the fact that demand is not a one-sided transaction, it makes nonsense to neglect it when we come to aggregate demand. This is the vital difference between the theory of the economic behaviour of the aggregate and the theory of the behaviour of the individual unit, in which we assume that changes in the individual's own demand do not affect his income.

BOOK III
THE PROPENSITY TO CONSUME

Chapter 8

THE PROPENSITY TO CONSUME:
I. THE OBJECTIVE FACTORS

I

We are now in a position to return to our main theme,
from which we broke off at the end of Book I in order
to deal with certain general problems of method and
definition. The ultimate object of our analysis is to
discover what determines the volume of employment.
So far we have established the preliminary conclusion
that the volume of employment is determined by the
point of intersection of the aggregate supply function
with the aggregate demand function. The aggregate
supply function, however, which depends in the main
on the physical conditions of supply, involves few considerations which are not already familiar. The form
may be unfamiliar but the underlying factors are not
new. We shall return to the aggregate supply function in chapter 20, where we discuss its inverse under
the name of the *employment function*. But, in the main,
it is the part played by the aggregate demand function
which has been overlooked; and it is to the aggregate
demand function that we shall devote Books III and IV.

The aggregate demand function relates any given
level of employment to the 'proceeds' which that level
of employment is expected to realise. The 'proceeds'
are made up of the sum of two quantities—the sum
which will be spent on consumption when employment
is at the given level, and the sum which will be devoted

to investment. The factors which govern these two quantities are largely distinct. In this book we shall consider the former, namely what factors determine the sum which will be spent on consumption when employment is at a given level; and in Book IV we shall proceed to the factors which determine the sum which will be devoted to investment.

Since we are here concerned in determining what sum will be spent on consumption when employment is at a given level, we should, strictly speaking, consider the function which relates the former quantity (C) to the latter (N). It is more convenient, however, to work in terms of a slightly different function, namely, the function which relates the consumption in terms of wage-units (C_w) to the income in terms of wage-units (Y_w) corresponding to a level of employment N. This suffers from the objection that Y_w is not a unique function of N, which is the same in all circumstances. For the relationship between Y_w and N may depend (though probably in a very minor degree) on the precise nature of the employment. That is to say, two different distributions of a given aggregate employment N between different employments might (owing to the different shapes of the individual employment functions—a matter to be discussed in Chapter 20 below) lead to different values of Y_w. In conceivable circumstances a special allowance might have to be made for this factor. But in general it is a good approximation to regard Y_w as uniquely determined by N. We will therefore define what we shall call *the propensity to consume* as the functional relationship χ between Y_w, a given level of income in terms of wage-units, and C_w the expenditure on consumption out of that level of income, so that

$$C_w = \chi(Y_w) \quad \text{or} \quad C = W \cdot \chi(Y_w).$$

The amount that the community spends on consumption obviously depends (i) partly on the amount of its income, (ii) partly on the other objective attendant circumstances, and (iii) partly on the subjective needs and the psychological propensities and habits of the individuals composing it and the principles on which the income is divided between them (which may suffer modification as output is increased). The motives to spending interact and the attempt to classify them runs the danger of false division. Nevertheless it will clear our minds to consider them separately under two broad heads which we shall call the subjective factors and the objective factors. The subjective factors, which we shall consider in more detail in the next chapter, include those psychological characteristics of human nature and those social practices and institutions which, though not unalterable, are unlikely to undergo a material change over a short period of time except in abnormal or revolutionary circumstances. In an historical enquiry or in comparing one social system with another of a different type, it is necessary to take account of the manner in which changes in the subjective factors may affect the propensity to consume. But, in general, we shall in what follows take the subjective factors as given; and we shall assume that the propensity to consume depends only on changes in the objective factors.

II

The principal objective factors which influence the propensity to consume appear to be the following:

(1) *A change in the wage-unit.* Consumption (C) is obviously much more a function of (in some sense) *real* income than of money-income. In a given state of technique and tastes and of social conditions determining the distribution of income, a man's real income will rise and fall with the amount of his command over labour-units, i.e. with the amount of his income measured in wage-units; though when the aggregate volume of output changes, his real income will (owing

to the operation of decreasing returns) rise less than in proportion to his income measured in wage-units. As a first approximation, therefore, we can reasonably assume that, if the wage-unit changes, the expenditure on consumption corresponding to a given level of employment will, like prices, change in the same proportion; though in some circumstances we may have to make an allowance for the possible reactions on aggregate consumption of the change in the distribution of a given real income between entrepreneurs and rentiers resulting from a change in the wage-unit. Apart from this, we have already allowed for changes in the wage-unit by defining the propensity to consume in terms of income measured in terms of wage-units.

(2) *A change in the difference between income and net income.* We have shown above that the amount of consumption depends on net income rather than on income, since it is, by definition, his net income that a man has primarily in mind when he is deciding his scale of consumption. In a given situation there may be a somewhat stable relationship between the two, in the sense that there will be a function uniquely relating different levels of income to the corresponding levels of net income. If, however, this should not be the case, such part of any change in income as is not reflected in net income must be neglected since it will have no effect on consumption; and, similarly, a change in net income, not reflected in income, must be allowed for. Save in exceptional circumstances, however, I doubt the practical importance of this factor. We will return to a fuller discussion of the effect on consumption of the difference between income and net income in the fourth section of this chapter.

(3) *Windfall changes in capital-values not allowed for in calculating net income.* These are of much more importance in modifying the propensity to consume, since they will bear no stable or regular relationship to the amount of income. The consumption of the

wealth-owning class may be extremely susceptible to unforeseen changes in the money-value of its wealth. This should be classified amongst the major factors capable of causing short-period changes in the propensity to consume.

(4) *Changes in the rate of time-discounting,* i.e. *in the ratio of exchange between present goods and future goods.* This is not quite the same thing as the rate of interest, since it allows for future changes in the purchasing power of money in so far as these are foreseen. Account has also to be taken of all kinds of risks, such as the prospect of not living to enjoy the future goods or of confiscatory taxation. As an approximation, however, we can identify this with the rate of interest.

The influence of this factor on the rate of spending out of a given income is open to a good deal of doubt. For the classical theory of the rate of interest,[1] which was based on the idea that the rate of interest was the factor which brought the supply and demand for savings into equilibrium, it was convenient to suppose that expenditure on consumption is *cet. par.* negatively sensitive to changes in the rate of interest, so that any rise in the rate of interest would appreciably diminish consumption. It has long been recognised, however, that the total effect of changes in the rate of interest on the readiness to spend on present consumption is complex and uncertain, being dependent on conflicting tendencies, since some of the subjective motives towards saving will be more easily satisfied if the rate of interest rises, whilst others will be weakened. Over a long period substantial changes in the rate of interest probably tend to modify social habits considerably, thus affecting the subjective propensity to spend—though in which direction it would be hard to say, except in the light of actual experience. The usual type of short-period fluctuation in the rate of interest is not likely, however, to have much *direct* influence on spending either way.

[1] Cf. chapter 14 below.

There are not many people who will alter their way of living because the rate of interest has fallen from 5 to 4 per cent, if their aggregate income is the same as before. Indirectly there may be more effects, though not all in the same direction. Perhaps the most important influence, operating through changes in the rate of interest, on the readiness to spend out of a given income, depends on the effect of these changes on the appreciation or depreciation in the price of securities and other assets. For if a man is enjoying a windfall increment in the value of his capital, it is natural that his motives towards current spending should be strengthened, even though in terms of income his capital is worth no more than before; and weakened if he is suffering capital losses. But this indirect influence we have allowed for already under (3) above. Apart from this, the main conclusion suggested by experience is, I think, that the short-period influence of the rate of interest on individual spending out of a given income is secondary and relatively unimportant, except, perhaps, where unusually large changes are in question. When the rate of interest falls very low indeed, the increase in the ratio between an annuity purchasable for a given sum and the annual interest on that sum may, however, provide an important source of negative saving by encouraging the practice of providing for old age by the purchase of an annuity.

The abnormal situation, where the propensity to consume may be sharply affected by the development of extreme uncertainty concerning the future and what it may bring forth, should also, perhaps, be classified under this heading.

(5) *Changes in fiscal policy.* In so far as the inducement to the individual to save depends on the future return which he expects, it clearly depends not only on the rate of interest but on the fiscal policy of the government. Income taxes, especially when they discriminate against 'unearned' income, taxes on

capital-profits, death-duties and the like are as relevant as the rate of interest; whilst the range of possible changes in fiscal policy may be greater, in expectation at least, than for the rate of interest itself. If fiscal policy is used as a deliberate instrument for the more equal distribution of incomes, its effect in increasing the propensity to consume is, of course, all the greater.[1]

We must also take account of the effect on the aggregate propensity to consume of government sinking funds for the discharge of debt paid for out of ordinary taxation. For these represent a species of corporate saving, so that a policy of substantial sinking funds must be regarded in given circumstances as reducing the propensity to consume. It is for this reason that a change-over from a policy of government borrowing to the opposite policy of providing sinking funds (or *vice versa*) is capable of causing a severe contraction (or marked expansion) of effective demand.

(6) *Changes in expectations of the relation between the present and the future level of income.* We must catalogue this factor for the sake of formal completeness. But, whilst it may affect considerably a particular individual's propensity to consume, it is likely to average out for the community as a whole. Moreover, it is a matter about which there is, as a rule, too much uncertainty for it to exert much influence.

We are left therefore, with the conclusion that in a given situation the propensity to consume may be considered a fairly stable function, provided that we have eliminated changes in the wage-unit in terms of money. Windfall changes in capital-values will be capable of changing the propensity to consume, and substantial

---

[1] It may be mentioned, in passing, that the effect of fiscal policy on the growth of wealth has been the subject of an important misunderstanding which, however, we cannot discuss adequately without the assistance of the theory of the rate of interest to be given in Book IV.

changes in the rate of interest and in fiscal policy may make some difference; but the other objective factors which might affect it, whilst they must not be overlooked, are not likely to be important in ordinary circumstances.

The fact that, given the general economic situation, the expenditure on consumption in terms of the wage-unit depends in the main, on the volume of output and employment is the justification for summing up the other factors in the portmanteau function 'propensity to consume'. For whilst the other factors are capable of varying (and this must not be forgotten), the aggregate income measured in terms of the wage-unit is, as a rule, the principal variable upon which the consumption-constituent of the aggregate demand function will depend.

III

Granted, then, that the propensity to consume is a fairly stable function so that, as a rule, the amount of aggregate consumption mainly depends on the amount of aggregate income (both measured in terms of wage-units), changes in the propensity itself being treated as a secondary influence, what is the normal shape of this function?

The fundamental psychological law, upon which we are entitled to depend with great confidence both *a priori* from our knowledge of human nature and from the detailed facts of experience, is that men are disposed, as a rule and on the average, to increase their consumption as their income increases, but not by as much as the increase in their income. That is to say, if C_w is the amount of consumption and Y_w is income (both measured in wage-units) ΔC_w has the same sign as ΔY_w but is smaller in amount, i.e. $\dfrac{dC_w}{dY_w}$ is positive and less than unity.

This is especially the case where we have short periods in view, as in the case of the so-called cyclical fluctuations of employment during which habits, as distinct from more permanent psychological propensities, are not given time enough to adapt themselves to changed objective circumstances. For a man's habitual standard of life usually has the first claim on his income, and he is apt to save the difference which discovers itself between his actual income and the expense of his habitual standard; or, if he does adjust his expenditure to changes in his income, he will over short periods do so imperfectly. Thus a rising income will often be accompanied by increased saving, and a falling income by decreased saving, on a greater scale at first than subsequently.

But, apart from short-period *changes* in the level of income, it is also obvious that a higher absolute level of income will tend, as a rule, to widen the gap between income and consumption. For the satisfaction of the immediate primary needs of a man and his family is usually a stronger motive than the motives towards accumulation, which only acquire effective sway when a margin of comfort has been attained. These reasons will lead, as a rule, to a *greater proportion* of income being saved as real income increases. But whether or not a greater proportion is saved, we take it as a fundamental psychological rule of any modern community that, when its real income is increased, it will not increase its consumption by an equal *absolute* amount, so that a greater absolute amount must be saved, unless a large and unusual change is occurring at the same time in other factors. As we shall show subsequently,[1] the stability of the economic system essentially depends on this rule prevailing in practice. This means that, if employment and hence aggregate income increase, *not all* the additional employment will be required to satisfy the needs of additional consumption.

---

[1] Cf. p. 251 below.

On the other hand, a decline in income due to a decline in the level of employment, if it goes far, may even cause consumption to exceed income not only by some individuals and institutions using up the financial reserves which they have accumulated in better times, but also by the government, which will be liable, willingly or unwillingly, to run into a budgetary deficit or will provide unemployment relief, for example, out of borrowed money. Thus, when employment falls to a low level, aggregate consumption will decline by a smaller amount than that by which real income has declined, by reason both of the habitual behaviour of individuals and also of the probable policy of governments; which is the explanation why a new position of equilibrium can usually be reached within a modest range of fluctuation. Otherwise a fall in employment and income, once started, might proceed to extreme lengths.

This simple principle leads, it will be seen, to the same conclusion as before, namely, that employment can only increase *pari passu* with an increase in investment; unless, indeed, there is a change in the propensity to consume. For since consumers will spend less than the increase in aggregate supply price when employment is increased, the increased employment will prove unprofitable unless there is an increase in investment to fill the gap.

IV

We must not underestimate the importance of the fact already mentioned above that, whereas employment is a function of the expected consumption and the expected investment, consumption is, *cet. par.*, a function of *net* income, i.e. of *net* investment (net income being equal to consumption *plus* net investment). In other words, the larger the financial provision which it is thought necessary to make before reckoning net income, the less favourable to consumption, and therefore to employment, will a given level of investment prove to be.

When the whole of this financial provision (or supplementary cost) is in fact currently expended in the upkeep of the already existing capital equipment, this point is not likely to be overlooked. But when the financial provision *exceeds* the actual expenditure on current upkeep, the practical results of this in its effect on employment are not always appreciated. For the amount of this excess neither directly gives rise to current investment nor is available to pay for consumption. It has, therefore, to be balanced by new investment, the demand for which has arisen quite independently of the current wastage of old equipment against which the financial provision is being made; with the result that the new investment available to provide current income is correspondingly diminished and a more intense demand for new investment is necessary to make possible a given level of employment. Moreover, much the same considerations apply to the allowance for wastage included in user cost, in so far as the wastage is not actually made good.

Take a house which continues to be habitable until it is demolished or abandoned. If a certain sum is written off its value out of the annual rent paid by the tenants, which the landlord neither spends on upkeep nor regards as net income available for consumption, this provision, whether it is a part of U or of V, constitutes a drag on employment all through the life of the house, suddenly made good in a lump when the house has to be rebuilt.

In a stationary economy all this might not be worth mentioning, since in each year the depreciation allowances in respect of old houses would be exactly offset by the new houses built in replacement of those reaching the end of their lives in that year. But such factors may be serious in a non-static economy, especially during a period which immediately succeeds a lively burst of investment in long-lived capital. For in such circumstances a very large proportion of the new items of investment may be absorbed by the larger financial provisions made by entrepreneurs in respect of existing capital equipment, upon the repairs and renewal of which, though it is wearing out with time, the date has not yet arrived for spending anything approaching the full financial provision which is being set aside; with the result that incomes cannot rise above a level which is low enough to correspond with a low aggregate of net investment. Thus sinking funds, etc., are apt to withdraw spending power from the consumer long before the demand for expenditure on replacements (which such provisions are anticipating) comes into play; i.e. they diminish the current effective demand and only increase it in the year in which the replacement is actually made. If the effect of this is aggravated by 'financial prudence', i.e. by its being thought advisable to 'write off' the initial cost *more* rapidly than the equipment actually wears out, the cumulative result may be very serious indeed.

In the United States, for example, by 1929 the rapid capital expansion of the previous five years had led cumulatively to the setting up of sinking funds and depreciation allowances, in respect of plant which did not need replacement, on so huge a scale that an enormous volume of entirely new investment was required merely to absorb these financial provisions; and it became almost hopeless to find still more new investment on a sufficient scale to provide for such new saving as a wealthy community in full employment would be disposed to set aside. This factor alone was probably sufficient to cause a slump. And, furthermore, since 'financial prudence' of this kind continued to be exercised through the slump by those great corporations which were still in a position to afford it, it offered a serious obstacle to early recovery.

Or again, in Great Britain at the present time (1935) the substantial amount of house-building and of other new investments since the war has led to an amount of sinking funds being set up much in excess of any present requirements for expenditure on repairs and renewals, a tendency which has been accentuated, where the investment has been made by local authorities and public boards, by the principles of 'sound' finance which often require sinking funds sufficient to write off the initial cost some time before replacement will actually fall due; with the result that even if private individuals were ready to spend the whole of their net incomes it would be a severe task to restore full employment in the face of this heavy volume of statutory provision by public and semi-public authorities, entirely dissociated from any corresponding new investment. The sinking funds of local authorities now stand, I think,[1] at an annual figure of more than half the amount which these authorities are spending on the whole of their new developments.[2] Yet it is not certain that the Ministry of Health are aware, when they insist on stiff sinking funds by local authorities, how much they may be aggravating the problem of unemployment. In the case of advances by building societies to help an individual to build his own house, the desire to be clear of debt more rapidly than the house actually deteriorates may stimulate the house-owner to save more than he otherwise would;—though this factor should be classified, perhaps, as diminishing the propensity to consume directly rather than through its effect on net income. In actual figures, repayments of mortgages advanced by building societies, which amounted to £24,000,000 in 1925, had risen to

[1] The actual figures are deemed of so little interest that they are only published two years or more in arrear.

[2] In the year ending March 31, 1930, local authorities spent £87,000,000 on capital account, of which £37,000,000 was provided by sinking funds, etc., in respect of previous capital expenditure; in the year ending 31 March 1933, the figures were £81,000,000 and £46,000,000.

£68,000,000 by 1933, as compared with new advances of £103,000,000; and to-day the repayments are probably still higher.

That it is investment, rather than net investment, which emerges from the statistics of output, is brought out forcibly and naturally in Mr Colin Clark's *National Income, 1924–1931*. He also shows what a large proportion depreciation, etc., normally bears to the value of investment. For example, he estimates that in Great Britain, over the years 1928–1931,[1] the investment and the net investment were as follows, though his gross investment is probably somewhat greater than my investment, inasmuch as it may include a part of user cost, and it is not clear how closely his 'net investment' corresponds to my definition of this term:

	(£ million)			
	1928	1929	1930	1931
Gross Investment–Output	791	731	620	482
'Value of physical wasting of old capital'	433	435	437	439
Net Investment	358	296	183	43

Mr Kuznets has arrived at much the same conclusion in compiling the statistics of the *Gross Capital Formation* (as he calls what I call investment) in the United States, 1919–1933. The physical fact, to which the statistics of output correspond, is inevitably the gross, and not the net, investment. Mr Kuznets has also discovered the difficulties in passing from gross investment to net investment. 'The difficulty', he writes, 'of passing from gross to net capital formation, that is, the difficulty of correcting for the consumption of existing durable commodities, is not only in the lack of data. The very concept of annual consumption of commodities that last over a number of years suffers from ambiguity'. He falls back, therefore, 'on the

[1] *Op. cit.* pp. 117 and 138.

assumption that the allowance for depreciation and depletion on the books of business firms describes correctly the volume of consumption of already existing, finished durable goods used by business firms'[1]. On the other hand, he attempts no deduction at all in respect of houses and other durable commodities in the hands of individuals. His very interesting results for the United States can be summarised as follows:[2]

Several facts emerge with prominence from this table. Net capital formation was very steady over the quinquennium 1925–1929, with only a 10 per cent.

	(Millions of dollars)				
	1925	1926	1927	1928	1929
Gross capital formation (after allowing for net change in business inventories)	30,706	33,571	31,157	33,934	34,491
Entrepreneurs' servicing, repairs, maintenance, depreciation and depletion	7,685	8,288	8,223	8,481	9,010
Net capital formation (on Mr Kuznets' definition)	23,021	25,283	22,934	25,453	25,481

	(Millions of dollars)			
	1930	1931	1932	1933
Gross capital formation (after allowing for net change in business inventories)	27,538	18,721	7,780	14,879
Entrepreneurs' servicing, repairs, maintenance, depreciation and depletion	8,502	7,623	6,543	8,204
Net capital formation (on Mr Kuznets' definition)	19,036	11,098	1,237	6,675

[1] These references are taken from a Bulletin (No. 52) of the National Bureau of Economic Research, giving preliminary results of Mr Kuznets' forthcoming book.
[2] See Appendix 2 below for further discussion of this point [Ed.].

increase in the latter part of the upward movement. The deduction for entrepreneurs' repairs, maintenance, depreciation and depletion remained at a high figure even at the bottom of the slump. But Mr Kuznets' method must surely lead to too low an estimate of the annual increase in depreciation, etc.; for he puts the latter at less than $1\frac{1}{2}$ per cent per annum of the new net capital formation. Above all, net capital formation suffered an appalling collapse after 1929, falling in 1932 to a figure no less than 95 per cent below the *average* of the quinquennium 1925–1929.

The above is, to some extent, a digression. But it is important to emphasise the magnitude of the deduction which has to be made from the income of a society, which already possesses a large stock of capital, before we arrive at the net income which is ordinarily available for consumption. For if we overlook this, we may underestimate the heavy drag on the propensity to consume which exists even in conditions where the public is ready to consume a very large proportion of its net income.

Consumption—to repeat the obvious—is the sole end and object of all economic activity. Opportunities for employment are necessarily limited by the extent of aggregate demand. Aggregate demand can be derived only from present consumption or from present provision for future consumption. The consumption for which we can profitably provide in advance cannot be pushed indefinitely into the future. We cannot, as a community, provide for future consumption by financial expedients but only by current physical output. In so far as our social and business organisation separates financial provision for the future from physical provision for the future so that efforts to secure the former do not necessarily carry the latter with them, financial prudence will be liable to diminish aggregate demand and thus impair well-being, as there are many examples to testify. The greater, moreover, the consumption for which we have provided in advance, the more difficult it is to find something further to provide for in advance, and the greater our dependence on present consumption as a source of demand. Yet the larger our incomes, the greater, unfortunately, is the margin between our incomes and our consumption. So, failing some novel expedient, there is, as we shall see, no answer to the riddle, except that there must be sufficient unemployment to keep us so poor that our consumption falls short of our income by no more than the equivalent of the physical provision for future consumption which it pays to produce to-day.

Or look at the matter thus. Consumption is satisfied partly by objects produced currently and partly by objects produced previously, i.e. by disinvestment. To the extent that consumption is satisfied by the latter, there is a contraction of current demand, since to that extent a part of current expenditure fails to find its way back as a part of net income. Contrariwise whenever an object is produced within the period with a view to satisfying consumption subsequently, an expansion of current demand is set up. Now all capital-investment is destined to result, sooner or later, in capital-disinvestment. Thus the problem of providing that new capital-investment shall always outrun capital-disinvestment sufficiently to fill the gap between net income and consumption, presents a problem which is increasingly difficult as capital increases. New capital-investment can only take place in excess of current capital-disinvestment if *future* expenditure on consumption is expected to increase. Each time we secure to-day's equilibrium by increased investment we are aggravating the difficulty of securing equilibrium to-morrow. A diminished propensity to consume to-day can only be accommodated to the public advantage if an increased propensity to consume is expected to exist some day. We are reminded of 'The Fable of the Bees'—the gay of to-morrow are absolutely indispensable to provide a *raison d'être* for the grave of to-day.

It is a curious thing, worthy of mention, that the popular mind seems only to be aware of this ultimate perplexity where *public* investment is concerned, as in the case of road-building and house-building and the like. It is commonly urged as an objection to schemes for raising employment by investment under the auspices of public authority that it is laying up trouble for the future. 'What will you do,' it is asked, 'when you have built all the houses and roads and town halls and electric grids and water supplies and so forth which the stationary population of the future can be expected to require?' But it is not so easily understood that the same difficulty applies to private investment and to industrial expansion; particularly to the latter, since it is much easier to see an early satiation of the demand for new factories and plant which absorb individually but little money, than of the demand for dwelling-houses.

The obstacle to a clear understanding is, in these examples, much the same as in many academic discussions of capital, namely, an inadequate appreciation of the fact that capital is not a self-subsistent entity existing apart from consumption. On the contrary, every weakening in the propensity to consume regarded as a permanent habit must weaken the demand for capital as well as the demand for consumption.

Chapter 9

THE PROPENSITY TO CONSUME:
II. THE SUBJECTIVE FACTORS

I

There remains the second category of factors which affect the amount of consumption out of a given income —namely, those subjective and social incentives which determine how much is spent, given the aggregate of income in terms of wage-units and given the relevant objective factors which we have already discussed. Since, however, the analysis of these factors raises no point of novelty, it may be sufficient if we give a catalogue of the more important, without enlarging on them at any length.

There are, in general, eight main motives or objects of a subjective character which lead individuals to refrain from spending out of their incomes:

(i) To build up a reserve against unforeseen contingencies;

(ii) To provide for an anticipated future relation between the income and the needs of the individual or his family different from that which exists in the present, as, for example, in relation to old age, family education, or the maintenance of dependents;

(iii) To enjoy interest and appreciation, i.e. because a larger real consumption at a later date is preferred to a smaller immediate consumption;

(iv) To enjoy a gradually increasing expenditure, since it gratifies a common instinct to look forward to a gradually improving standard of life rather than the contrary, even though the capacity for enjoyment may be diminishing;

(v)   To enjoy a sense of independence and the power to do things, though without a clear idea or definite intention of specific action;

(vi) To secure a *masse de manœuvre* to carry out speculative or business projects;

(vii) To bequeath a fortune;

(viii) To satisfy pure miserliness, i.e. unreasonable but insistent inhibitions against acts of expenditure as such.

These eight motives might be called the motives of Precaution, Foresight, Calculation, Improvement, Independence, Enterprise, Pride and Avarice; and we could also draw up a corresponding list of motives to consumption such as Enjoyment, Shortsightedness, Generosity, Miscalculation, Ostentation and Extravagance.

Apart from the savings accumulated by individuals, there is also the large amount of income, varying perhaps from one-third to two-thirds of the total accumulation in a modern industrial community such as Great Britain or the United States, which is withheld by central and local government, by institutions and by business corporations—for motives largely analogous to, but not identical with, those actuating individuals, and mainly the four following:

(i) The motive of enterprise—to secure resources to carry out further capital investment without incurring debt or raising further capital on the market;

(ii) The motive of liquidity—to secure liquid resources to meet emergencies, difficulties and depressions;

(iii) The motive of improvement—to secure a gradually increasing income, which, incidentally, will protect the management from criticism, since increasing income due to accumulation is seldom distinguished from increasing income due to efficiency;

(iv) The motive of financial prudence and the anxiety to be 'on the right side' by making a financial provision in excess of user and supplementary cost, so as to discharge debt and write off the cost of assets ahead of, rather than behind, the actual rate of wastage and obsolescence, the strength of this motive mainly depending on the quantity and character of the capital equipment and the rate of technical change.

Corresponding to these motives which favour the withholding of a part of income from consumption, there are also operative at times motives which lead to an excess of consumption over income. Several of the motives towards positive saving catalogued above as affecting individuals have their intended counterpart in negative saving at a later date, as, for example, with saving to provide for family needs or old age. Unemployment relief financed by borrowing is best regarded as negative saving.

Now the strength of all these motives will vary enormously according to the institutions and organisation of the economic society which we presume, according to habits formed by race, education, convention, religion and current morals, according to present hopes and past experience, according to the scale and technique of capital equipment, and according to the prevailing distribution of wealth and the established standards of life. In the argument of this book, however, we shall not concern ourselves, except in occasional digressions, with the results of far-reaching social changes or with the slow effects of secular progress. We shall, that is to say, take as given the main background of subjective motives to saving and to consumption respectively. In so far as the distribution of wealth is determined by the more or less permanent social structure of the community, this also can be reckoned a factor, subject only to slow change and over a long period, which we can take as given in our present context.

II

Since, therefore, the main background of subjective and social incentives changes slowly, whilst the short-period influence of changes in the rate of interest and the other objective factors is often of secondary importance, we are left with the conclusion that short-period changes in consumption largely depend on changes in the rate at which income (measured in wage-units) is being earned and not on changes in the propensity to consume out of a given income.

We must, however, guard against a misunderstanding. The above means that the influence of moderate changes in the rate of interest on the *propensity* to consume is usually small. It does not mean that changes in the rate of interest have only a small influence on the amounts *actually* saved and consumed. Quite the contrary. The influence of changes in the rate of interest on the amount actually saved is of paramount importance, but is *in the opposite direction* to that usually supposed. For even if the attraction of the larger future income to be earned from a higher rate of interest has the effect of diminishing the propensity to consume, nevertheless we can be certain that a rise in the rate of interest will have the effect of reducing the amount actually saved. For aggregate saving is governed by aggregate investment; a rise in the rate of interest (unless it is offset by a corresponding change in the demand-schedule for investment) will diminish investment; hence a rise in the rate of interest must have the effect of reducing incomes to a level at which saving is decreased in the same measure as investment. Since incomes will decrease by a greater absolute amount than investment, it is, indeed, true that, when the rate of interest rises, the rate of consumption will decrease. But this does not mean that there will be a wider margin for saving. On the contrary, saving and spending will *both* decrease.

Thus, even if it is the case that a rise in the rate of interest would cause the community to save more *out of a given income*, we can be quite sure that a rise in the rate of interest (assuming no favourable change in the demand-schedule for investment) will decrease the actual aggregate of savings. The same line of argument can even tell us by how much a rise in the rate of interest will, *cet. par.*, decrease incomes. For incomes will have to fall (or be redistributed) by just that amount which is required, with the existing propensity to consume, to decrease savings by the same amount by which the rise in the rate of interest will, with the existing marginal efficiency of capital, decrease investment. A detailed examination of this aspect will occupy our next chapter.

The rise in the rate of interest might induce us to save more, *if* our incomes were unchanged. But if the higher rate of interest retards investment, our incomes will not, and cannot, be unchanged. They must necessarily fall, until the declining capacity to save has sufficiently offset the stimulus to save given by the higher rate of interest. The more virtuous we are, the more determinedly thrifty, the more obstinately orthodox in our national and personal finance, the more our incomes will have to fall when interest rises relatively to the marginal efficiency of capital. Obstinacy can bring only a penalty and no reward. For the result is inevitable.

Thus, after all, the actual rates of aggregate saving and spending do not depend on Precaution, Foresight, Calculation, Improvement, Independence, Enterprise, Pride or Avarice. Virtue and vice play no part. It all depends on how far the rate of interest is favourable to investment, after taking account of the marginal efficiency of capital.[1] No, this is an overstatement. If the rate of interest were so governed as to maintain continuous full employment, virtue would resume her sway;—the rate of capital accumulation would depend on the weakness of the propensity to consume. Thus, once again, the tribute that classical economists pay to her is due to their concealed assumption that the rate of interest always is so governed.

[1] In some passages of this section we have tacitly anticipated ideas which will be introduced in Book IV.

Chapter 10

THE MARGINAL PROPENSITY TO CONSUME AND THE MULTIPLIER

We established in chapter 8 that employment can only increase *pari passu* with investment unless there is a change in the propensity to consume. We can now carry this line of thought a stage further. For in given circumstances a definite ratio, to be called the *multiplier*, can be established between income and investment and, subject to certain simplifications, between the total employment and the employment directly employed on investment (which we shall call the *primary employment*). This further step is an integral part of our theory of employment, since it establishes a precise relationship, given the propensity to consume, between aggregate employment and income and the rate of investment. The conception of the multiplier was first introduced into economic theory by Mr R. F. Kahn in his article on 'The Relation of Home Investment to Unemployment' (*Economic Journal*, June 1931). His argument in this article depended on the fundamental notion that, if the propensity to consume in various hypothetical circumstances is (together with certain other conditions) taken as given and we conceive the monetary or other public authority to take steps to stimulate or to retard investment, the change in the amount of employment will be a function of the net change in the amount of investment; and it aimed at laying down general principles by which to estimate the actual quantitative relationship between an increment of net investment and the increment of aggregate employment which will be associated with it. Before coming to the multiplier, however, it will be convenient to introduce the conception of the *marginal propensity to consume*.

I

The fluctuations in real income under consideration in this book are those which result from applying different quantities of employment (i.e. of labour-units) to a given capital equipment, so that real income increases and decreases with the number of labour-units employed. If, as we assume in general, there is a decreasing return at the margin as the number of labour-units employed on the given capital equipment is increased, income measured in terms of wage-units will increase more than in proportion to the amount of employment, which, in turn, will increase more than in proportion to the amount of real income measured (if that is possible) in terms of product. Real income measured in terms of product and income measured in terms of wage-units will, however, increase and decrease together (in the short period when capital equipment is virtually unchanged). Since, therefore, real income, in terms of product, may be incapable of precise numerical measurement, it is often convenient to regard income in terms of wage-units (Y_w) as an adequate working index of changes in real income. In certain contexts we must not overlook the fact that, in general, Y_w increases and decreases in a greater proportion than real income; but in other contexts the fact that they always increase and decrease together renders them virtually interchangeable.

Our normal psychological law that, when the real income of the community increases or decreases, its consumption will increase or decrease but not so fast, can, therefore, be translated—not, indeed, with absolute accuracy but subject to qualifications which are obvious and can easily be stated in a formally complete fashion—into the propositions that ΔC_w and ΔY_w have the same sign, but $\Delta Y_w > \Delta C_w$, where C_w is the consumption in terms of wage-units. This is merely a repetition of the proposition already established on p. 29 above.

Let us define, then, $\dfrac{dC_w}{dY_w}$ as the *marginal propensity to consume*.

This quantity is of considerable importance, because it tells us how the next increment of output will have to be divided between consumption and investment. For $\Delta Y_w = \Delta C_w + \Delta I_w$, where ΔC_w and ΔI_w are the increments of consumption and investment; so that we can write $\Delta Y_w = k\Delta I_w$, where $1 - \dfrac{1}{k}$ is equal to the marginal propensity to consume.

Let us call k the *investment multiplier*. It tells us that, when there is an increment of aggregate investment, income will increase by an amount which is k times the increment of investment.

II

Mr Kahn's multiplier is a little different from this, being what we may call the *employment multiplier* designated by k', since it measures the ratio of the increment of total employment which is associated with a given increment of primary employment in the investment industries. That is to say, if the increment of investment ΔI_w leads to an increment of primary employment ΔN_2 in the investment industries, the increment of total employment $\Delta N = k'\Delta N_2$.

There is no reason in general to suppose that $k = k'$. For there is no necessary presumption that the shapes of the relevant portions of the aggregate supply functions for different types of industry are such that the ratio of the increment of employment in the one set of industries to the increment of demand which has stimulated it will be the same as in the other set of industries.[1] It is easy, indeed, to conceive of cases, as, for example, where the marginal propensity to consume is widely different from the average propensity, in which there would be a presumption in favour of some inequality between $\dfrac{\Delta Y_w}{\Delta N}$ and $\dfrac{\Delta I_w}{\Delta N_2}$, since there would be very divergent proportionate changes in the demands for consumption-goods and investment-goods respectively. If we wish to take account of such possible differences in the shapes of the relevant portions of the aggregate supply functions for the two groups of industries respectively, there is no difficulty in rewriting the following argument in the more generalised form. But to elucidate the ideas involved, it will be convenient to deal with the simplified case where $k = k'$.

It follows, therefore, that, if the consumption psychology of the community is such that they will choose to consume, e.g. nine-tenths of an increment of income,[2] then the multiplier k is 10; and the total employment caused by (e.g.) increased public works will be ten times the primary employment provided by

[1] More precisely, if e_e and e'_e are the elasticities of employment in industry as a whole and in the investment industries respectively, and if N and N_2 are the numbers of men employed in industry as a whole and in the investment industries, we have

$$\Delta Y_w = \frac{Y_w}{e_e . N}\Delta N$$

and

$$\Delta I_w = \frac{I_w}{e'_e . N_2}\Delta N_2,$$

so that

$$\Delta N = \frac{e_e I_w N}{e'_e N_2 Y_w}k . \Delta N_2,$$

i.e.

$$k' = \frac{I_w}{e'_e N_2} . \frac{e_e N}{Y_w}k.$$

If, however, there is no reason to expect any material relevant difference in the shapes of the aggregate supply functions for industry as a whole and for the investment industries respectively, so that $\dfrac{I_w}{e'_e . N_2} = \dfrac{Y_w}{e_e . N}$, then it follows that $\dfrac{\Delta Y_w}{\Delta N} = \dfrac{\Delta I_w}{\Delta N_2}$ and, therefore, that $k = k'$.

[2] Our quantities are measured throughout in terms of wage-units.

the public works themselves, assuming no reduction of investment in other directions. Only in the event of the community maintaining their consumption unchanged in spite of the increase in employment and hence in real income, will the increase of employment be restricted to the primary employment provided by the public works. If, on the other hand, they seek to consume the whole of any increment of income, there will be no point of stability and prices will rise without limit. With normal psychological suppositions, an increase in employment will only be associated with a decline in consumption if there is at the same time a change in the propensity to consume—as the result, for instance, of propaganda in time of war in favour of restricting individual consumption; and it is only in this event that the increased employment in investment will be associated with an unfavourable repercussion on employment in the industries producing for consumption.

This only sums up in a formula what should by now be obvious to the reader on general grounds. An increment of investment in terms of wage-units cannot occur unless the public are prepared to increase their savings in terms of wage-units. Ordinarily speaking, the public will not do this unless their aggregate income in terms of wage-units is increasing. Thus their effort to consume a part of their increased incomes will stimulate output until the new level (and distribution) of incomes provides a margin of saving sufficient to correspond to the increased investment. The multiplier tells us by how much their employment has to be increased to yield an increase in real income sufficient to induce them to do the necessary extra saving, and is a function of their psychological propensities.[1] If saving is the pill and consumption is the jam, the extra jam has to be proportioned to the size of the

[1] Though in the more generalised case it is also a function of the physical conditions of production in the investment and consumption industries respectively.

additional pill. Unless the psychological propensities of the public are different from what we are supposing, we have here established the law that increased employment for investment must necessarily stimulate the industries producing for consumption and thus lead to a total increase of employment which is a multiple of the primary employment required by the investment itself.

It follows from the above that, if the marginal propensity to consume is not far short of unity, small fluctuations in investment will lead to wide fluctuations in employment; but, at the same time, a comparatively small increment of investment will lead to full employment. If, on the other hand, the marginal propensity to consume is not much above zero, small fluctuations in investment will lead to correspondingly small fluctuations in employment; but, at the same time, it may require a large increment of investment to produce full employment. In the former case involuntary unemployment would be an easily remedied malady, though liable to be troublesome if it is allowed to develop. In the latter case, employment may be less variable but liable to settle down at a low level and to prove recalcitrant to any but the most drastic remedies. In actual fact the marginal propensity to consume seems to lie somewhere between these two extremes, though much nearer to unity than to zero; with the result that we have, in a sense, the worst of both worlds, fluctuations in employment being considerable and, at the same time, the increment in investment required to produce full employment being too great to be easily handled. Unfortunately the fluctuations have been sufficient to prevent the nature of the malady from being obvious, whilst its severity is such that it cannot be remedied unless its nature is understood.

When full employment is reached, any attempt to increase investment still further will set up a tendency in money-prices to rise without limit, irrespective of the marginal propensity to consume; i.e. we shall have reached a state of true inflation.[1] Up to this point, however, rising prices will be associated with an increasing aggregate real income.

III

We have been dealing so far with a *net* increment of investment. If, therefore, we wish to apply the above without qualification to the effect of (e.g.) increased public works, we have to assume that there is no offset through decreased investment in other directions,—and also, of course, no associated change in the propensity of the community to consume. Mr Kahn was mainly concerned in the article referred to above in considering what offsets we ought to take into account as likely to be important, and in suggesting quantitative estimates. For in an actual case there are several factors besides some specific increase of investment of a given kind which enter into the final result. If, for example, a government employs 100,000 additional men on public works, and if the multiplier (as defined above) is 4, it is not safe to assume that aggregate employment will increase by 400,000. For the new policy may have adverse reactions on investment in other directions.

It would seem (following Mr Kahn) that the following are likely in a modern community to be the factors which it is most important not to overlook (though the first two will not be fully intelligible until after Book IV has been reached):

(i) The method of financing the policy and the increased working cash, required by the increased employment and the associated rise of prices, may have the effect of increasing the rate of interest and so retarding investment in other directions, unless the monetary authority takes steps to the contrary; whilst, at the same time, the increased cost of capital goods will reduce their marginal efficiency to the private in-

[1] Cf. chapter 21, p. 303, below.

vestor, and this will require an actual *fall* in the rate of interest to offset it.

(ii) With the confused psychology which often prevails, the government programme may, through its effect on 'confidence', increase liquidity-preference or diminish the marginal efficiency of capital, which, again, may retard other investment unless measures are taken to offset it.

(iii) In an open system with foreign-trade relations, some part of the multiplier of the increased investment will accrue to the benefit of employment in foreign countries, since a proportion of the increased consumption will diminish our own country's favourable foreign balance; so that, if we consider only the effect on domestic employment as distinct from world employment, we must diminish the full figure of the multiplier. On the other hand our own country may recover a portion of this leakage through favourable repercussions due to the action of the multiplier in the foreign country in increasing its economic activity.

Furthermore, if we are considering changes of a substantial amount, we have to allow for a progressive change in the marginal propensity to consume, as the position of the margin is gradually shifted; and hence in the multiplier. The marginal propensity to consume is not constant for all levels of employment, and it is probable that there will be, as a rule, a tendency for it to diminish as employment increases; when real income increases, that is to say, the community will wish to consume a gradually diminishing proportion of it.

There are also other factors, over and above the operation of the general rule just mentioned, which may operate to modify the marginal propensity to consume, and hence the multiplier; and these other factors seem likely, as a rule, to accentuate the tendency of the general rule rather than to offset it. For, in the first place, the increase of employment will tend, owing to the effect of diminishing returns in the short period, to increase the proportion of aggregate income which accrues to the entrepreneurs, whose individual marginal propensity to consume is probably less than the average for the community as a whole. In the second place, unemployment is likely to be associated with negative saving in certain quarters, private or public, because the unemployed may be living either on the savings of themselves and their friends or on public relief which is partly financed out of loans; with the result that re-employment will gradually diminish these particular acts of negative saving and reduce, therefore, the marginal propensity to consume more rapidly than would have occurred from an equal increase in the community's real income accruing in different circumstances.

In any case, the multiplier is likely to be greater for a small net increment of investment than for a large increment; so that, where substantial changes are in view, we must be guided by the average value of the multiplier based on the average marginal propensity to consume over the range in question.

Mr Kahn has examined the probable quantitative result of such factors as these in certain hypothetical special cases. But, clearly, it is not possible to carry any generalisation very far. One can only say, for example, that a typical modern community would probably tend to consume not much less than 80 per cent of any increment of real income, if it were a closed system with the consumption of the unemployed paid for by transfers from the consumption of other consumers, so that the multiplier after allowing for offsets would not be much less than 5. In a country, however, where foreign trade accounts for, say, 20 per cent of consumption and where the unemployed receive out of loans or their equivalent up to, say, 50 per cent of their normal consumption when in work, the multiplier may fall as low as 2 or 3 times the employment provided by a specific new investment. Thus a given fluctuation of investment will be associated with a much less violent fluctuation of employment in a country in which foreign trade plays a large part and unemployment relief is financed on a larger scale out of borrowing (as was the case, e.g. in Great Britain in 1931), than in a country in which these factors are less important (as in the United States in 1932).[1]

It is, however, to the general principle of the multiplier to which we have to look for an explanation of how fluctuations in the amount of investment, which are a comparatively small proportion of the national income, are capable of generating fluctuations in aggregate employment and income so much greater in amplitude than themselves.

IV

The discussion has been carried on, so far, on the basis of a change in aggregate investment which has been foreseen sufficiently in advance for the consumption industries to advance *pari passu* with the capital-goods industries without more disturbance to the price of consumption-goods than is consequential, in conditions of decreasing returns, on an increase in the quantity which is produced.

In general, however, we have to take account of the case where the initiative comes from an increase in the output of the capital-goods industries which was not fully foreseen. It is obvious that an initiative of this description only produces its full effect on employment over a period of time. I have found, however, in discussion that this obvious fact often gives rise to some confusion between the logical theory of the multiplier, which holds good continuously, without time-lag, at all moments of time, and the consequences of an expansion in the capital-goods industries which take gradual effect, subject to time-lag and only after an interval.

[1] Cf. however, below, p. 128, for an American estimate.

The relationship between these two things can be cleared up by pointing out, firstly that an unforeseen, or imperfectly foreseen, expansion in the capital-goods industries does not have an instantaneous effect of equal amount on the aggregate of investment but causes a gradual increase of the latter; and, secondly, that it may cause a temporary departure of the marginal propensity to consume away from its normal value, followed, however, by a gradual return to it.

Thus an expansion in the capital-goods industries causes a series of increments in aggregate investment occurring in successive periods over an interval of time, and a series of values of the marginal propensity to consume in these successive periods which differ both from what the values would have been if the expansion had been foreseen and from what they will be when the community has settled down to a new steady level of aggregate investment. But in every interval of time the theory of the multiplier holds good in the sense that the increment of aggregate demand is equal to the product of the increment of aggregate investment and the multiplier as determined by the marginal propensity to consume.

The explanation of these two sets of facts can be seen most clearly by taking the extreme case where the expansion of employment in the capital-goods industries is so entirely unforeseen that in the first instance there is no increase whatever in the output of consumption-goods. In this event the efforts of those newly employed in the capital-goods industries to consume a proportion of their increased incomes will raise the prices of consumption-goods until a temporary equilibrium between demand and supply has been brought about partly by the high prices causing a postponement of consumption, partly by a redistribution of income in favour of the saving classes as an effect of the increased profits resulting from the higher prices, and partly by the higher prices causing a depletion

of stocks. So far as the balance is restored by a postponement of consumption there is a temporary reduction of the marginal propensity to consume, i.e.
of the multiplier itself, and in so far as there is a depletion of stocks, aggregate investment increases for
the time being by less than the increment of investment
in the capital-goods industries,—i.e. the thing to be
multiplied does not increase by the full increment of
investment in the capital-goods industries. As time
goes on, however, the consumption-goods industries
adjust themselves to the new demand, so that when the
deferred consumption is enjoyed, the marginal propensity to consume rises temporarily above its normal
level, to compensate for the extent to which it previously fell below it, and eventually returns to its normal
level; whilst the restoration of stocks to their previous
figure causes the increment of aggregate investment
to be temporarily greater than the increment of investment in the capital-goods industries (the increment of
working capital corresponding to the greater output
also having temporarily the same effect).

The fact that an unforeseen change only exercises
its full effect on employment over a period of time is
important in certain contexts;—in particular it plays
a part in the analysis of the trade cycle (on lines such
as I followed in my *Treatise on Money*). But it does not
in any way affect the significance of the theory of the
multiplier as set forth in this chapter; nor render it
inapplicable as an indicator of the total benefit to employment to be expected from an expansion in the
capital goods industries. Moreover, except in conditions where the consumption industries are already
working almost at capacity so that an expansion of output requires an expansion of plant and not merely the
more intensive employment of the existing plant, there
is no reason to suppose that more than a brief interval
of time need elapse before employment in the consumption industries is advancing *pari passu* with

employment in the capital-goods industries with the multiplier operating near its normal figure.

V

We have seen above that the greater the marginal propensity to consume, the greater the multiplier, and hence the greater the disturbance to employment corresponding to a given change in investment. This might seem to lead to the paradoxical conclusion that a poor community in which saving is a very small proportion of income will be more subject to violent fluctuations than a wealthy community where saving is a larger proportion of income and the multiplier consequently smaller.

This conclusion, however, would overlook the distinction between the effects of the marginal propensity to consume and those of the average propensity to consume. For whilst a high marginal propensity to consume involves a larger *proportionate* effect from a given percentage change in investment, the *absolute* effect will, nevertheless, be small if the *average* propensity to consume is also high. This may be illustrated as follows by a numerical example.

Let us suppose that a community's propensity to consume is such that, so long as its real income does not exceed the output from employing 5,000,000 men on its existing capital equipment, it consumes the whole of its income; that of the output of the next 100,000 additional men employed it consumes 99 per cent, of the next 100,000 after that 98 per cent, of the third 100,000 97 per cent and so on; and that 10,000,000 men employed represents full employment. It follows from this that, when $5,000,000 + n \times 100,000$ men are employed, the multiplier at the margin is $\dfrac{100}{n}$, and $\dfrac{n(n+1)}{2(50+n)}$ per cent of the national income is invested.

Thus when 5,200,000 men are employed the multiplier is very large, namely 50, but investment is only a trifling proportion of current income, namely, 0·06 per cent; with the result that if investment falls off by a large proportion, say about two-thirds, employment will only decline to 5,100,000, i.e. by about 2 per cent. On the other hand, when 9,000,000 men are employed, the marginal multiplier is comparatively small, namely $2\frac{1}{2}$, but investment is now a substantial proportion of current income, namely, 9 per cent; with the result that if investment falls by two-thirds, employment will decline to 6,900,000, namely, by 19 per cent. In the limit where investment falls off to zero, employment will decline by about 4 per cent in the former case, whereas in the latter case it will decline by 44 per cent.[1]

In the above example, the poorer of the two communities under comparison is poorer by reason of under-employment. But the same reasoning applies by easy adaptation if the poverty is due to inferior skill, technique or equipment. Thus whilst the multiplier is larger in a poor community, the effect on employment of fluctuations in investment will be much greater in a wealthy community, assuming that in the latter current investment represents a much larger proportion of current output.[2]

[1] Quantity of investment is measured, above, by the number of men employed in producing it. Thus if there are diminishing returns per unit of employment as employment increases, what is double the quantity of investment on the above scale will be less than double on a physical scale (if such a scale is available).

[2] More generally, the ratio of the proportional change in total demand to the proportional change in investment

$$= \frac{\Delta Y}{Y} \bigg/ \frac{\Delta I}{I} = \frac{\Delta Y}{Y} \cdot \frac{Y-C}{\Delta Y-\Delta C} = \frac{1-\dfrac{C}{Y}}{1-\dfrac{dC}{dY}}.$$

As wealth increases $\dfrac{dC}{dY}$ diminishes, but $\dfrac{C}{Y}$ also diminishes. Thus the fraction increases or diminishes according as consumption increases or diminishes in a smaller or greater proportion than income.

It is also obvious from the above that the employment of a given number of men on public works will (on the assumptions made) have a much larger effect on aggregate employment at a time when there is severe unemployment, than it will have later on when full employment is approached. In the above example, if, at a time when employment has fallen to 5,200,000, an additional 100,000 men are employed on public works, total employment will rise to 6,400,000. But if employment is already 9,000,000 when the additional 100,000 men are taken on for public works, total employment will only rise to 9,200,000. Thus public works even of doubtful utility may pay for themselves over and over again at a time of severe unemployment, if only from the diminished cost of relief expenditure, provided that we can assume that a smaller proportion of income is saved when unemployment is greater; but they may become a more doubtful proposition as a state of full employment is approached. Furthermore, if our assumption is correct that the marginal propensity to consume falls off steadily as we approach full employment, it follows that it will become more and more troublesome to secure a further given increase of employment by further increasing investment.

It should not be difficult to compile a chart of the marginal propensity to consume at each stage of a trade cycle from the statistics (if they were available) of aggregate income and aggregate investment at successive dates. At present, however, our statistics are not accurate enough (or compiled sufficiently with this specific object in view) to allow us to infer more than highly approximate estimates. The best for the purpose, of which I am aware, are Mr Kuznets' figures for the United States (already referred to, p. 103 above), though they are, nevertheless, very precarious. Taken in conjunction with estimates of national income these suggest, for what they are worth, both a lower figure and a more stable figure for the investment multiplier than I should have expected. If single years are taken in isolation, the results look rather wild. But if they are grouped in pairs, the multiplier seems to have been less than 3 and probably fairly stable in the neighbourhood of 2·5. This suggests a marginal propensity to consume not exceeding 60 to 70 per cent—a figure quite plausible for the boom, but surprisingly, and, in my judgment, improbably low for the slump. It is possible, however, that the extreme financial conservatism of corporate finance in the United States, even during the slump, may account for it. In other words, if, when investment is falling heavily through a failure to undertake repairs and replacements, financial provision is made, nevertheless, in respect of such wastage, the effect is to prevent the rise in the marginal propensity to consume which would have occurred otherwise. I suspect that this factor may have played a significant part in aggravating the degree of the recent slump in the United States. On the other hand, it is possible that the statistics somewhat overstate the decline in investment, which is alleged to have fallen off by more than 75 per cent in 1932 compared with 1929, whilst net 'capital formation' declined by more than 95 per cent;—a moderate change in these estimates being capable of making a substantial difference to the multiplier.

VI

When involuntary unemployment exists, the marginal disutility of labour is necessarily less than the utility of the marginal product. Indeed it may be much less. For a man who has been long unemployed some measure of labour, instead of involving disutility, may have a positive utility. If this is accepted, the above reasoning shows how 'wasteful' loan expenditure[1] may nevertheless enrich the com-

---

[1] It is often convenient to use the term 'loan expenditure' to include the public investment financed by borrowing from individuals and also

munity on balance. Pyramid-building, earthquakes, even wars may serve to increase wealth, if the education of our statesmen on the principles of the classical economics stands in the way of anything better.

It is curious how common sense, wriggling for an escape from absurd conclusions, has been apt to reach a preference for *wholly* 'wasteful' forms of loan expenditure rather than for *partly* wasteful forms, which, because they are not wholly wasteful, tend to be judged on strict 'business' principles. For example, unemployment relief financed by loans is more readily accepted than the financing of improvements at a charge below the current rate of interest; whilst the form of digging holes in the ground known as gold-mining, which not only adds nothing whatever to the real wealth of the world but involves the disutility of labour, is the most acceptable of all solutions.

If the Treasury were to fill old bottles with banknotes, bury them at suitable depths in disused coalmines which are then filled up to the surface with town rubbish, and leave it to private enterprise on well-tried principles of *laissez-faire* to dig the notes up again (the right to do so being obtained, of course, by tendering for leases of the note-bearing territory), there need be no more unemployment and, with the help of the repercussions, the real income of the community, and its capital wealth also, would probably become a good deal greater than it actually is. It would, indeed, be more sensible to build houses and the like; but if there are political and practical difficulties in the way of this, the above would be better than nothing.

The analogy between this expedient and the gold-

any other current public expenditure which is so financed. Strictly speaking, the latter should be reckoned as negative saving, but official action of this kind is not influenced by the same sort of psychological motives as those which govern private saving. Thus 'loan expenditure' is a convenient expression for the net borrowings of public authorities on all accounts, whether on capital account or to meet a budgetary deficit. The one form of loan expenditure operates by increasing investment and the other by increasing the propensity to consume.

mines of the real world is complete. At periods when gold is available at suitable depths experience shows that the real wealth of the world increases rapidly; and when but little of it is so available our wealth suffers stagnation or decline. Thus gold-mines are of the greatest value and importance to civilisation. Just as wars have been the only form of large-scale loan expenditure which statesmen have thought justifiable, so gold-mining is the only pretext for digging holes in the ground which has recommended itself to bankers as sound finance; and each of these activities has played its part in progress—failing something better. To mention a detail, the tendency in slumps for the price of gold to rise in terms of labour and materials aids eventual recovery, because it increases the depth at which gold-digging pays and lowers the minimum grade of ore which is payable.

In addition to the probable effect of increased supplies of gold on the rate of interest, gold-mining is for two reasons a highly practical form of investment, if we are precluded from increasing employment by means which at the same time increase our stock of useful wealth. In the first place, owing to the gambling attractions which it offers it is carried on without too close a regard to the ruling rate of interest. In the second place the result, namely, the increased stock of gold, does not, as in other cases, have the effect of diminishing its marginal utility. Since the value of a house depends on its utility, every house which is built serves to diminish the prospective rents obtainable from further house-building and therefore lessens the attraction of further similar investment unless the rate of interest is falling *pari passu*. But the fruits of gold-mining do not suffer from this disadvantage, and a check can only come through a rise of the wage-unit in terms of gold, which is not likely to occur unless and until employment is substantially better. Moreover, there is no subsequent reverse effect on account of provision for user and supplementary costs, as in the case of less durable forms of wealth.

Ancient Egypt was doubly fortunate, and doubtless owed to this its fabled wealth, in that it possessed *two* activities, namely, pyramid-building as well as the search for the precious metals, the fruits of which, since they could not serve the needs of man by being consumed, did not stale with abundance. The Middle Ages built cathedrals and sang dirges. Two pyramids, two masses for the dead, are twice as good as one; but not so two railways from London to York. Thus we are so sensible, have schooled ourselves to so close a semblance of prudent financiers, taking careful thought before we add to the 'financial' burdens of posterity by building them houses to live in, that we have no such easy escape from the sufferings of unemployment. We have to accept them as an inevitable result of applying to the conduct of the State the maxims which are best calculated to 'enrich' an individual by enabling him to pile up claims to enjoyment which he does not intend to exercise at any definite time.

BOOK IV

THE INDUCEMENT TO INVEST

Chapter 11

THE MARGINAL EFFICIENCY
OF CAPITAL

I

When a man buys an investment or capital-asset, he purchases the right to the series of prospective returns, which he expects to obtain from selling its output, after deducting the running expenses of obtaining that output, during the life of the asset. This series of annuities Q_1, Q_2, ... Q_n it is convenient to call the *prospective yield* of the investment.

Over against the prospective yield of the investment we have the *supply price* of the capital-asset, meaning by this, not the market-price at which an asset of the type in question can actually be purchased in the market, but the price which would just induce a manufacturer newly to produce an additional unit of such assets, i.e. what is sometimes called its *replacement cost*. The relation between the prospective yield of a capital-asset and its supply price or replacement cost, i.e. the relation between the prospective yield of one more unit of that type of capital and the cost of producing that unit, furnishes us with the *marginal efficiency of capital* of that type. More precisely, I define the marginal efficiency of capital as being equal to that rate of discount which would make the present value of the series of annuities given by the returns expected from the capital-asset during its life just equal to its supply price. This gives us the marginal efficiencies of particular types of capital-assets. The greatest of these marginal efficiencies can then be regarded as the marginal efficiency of capital in general.

The reader should note that the marginal efficiency of capital is here defined in terms of the *expectation* of yield and of the *current* supply price of the capital-asset. It depends on the rate of return expected to be obtainable on money if it were invested in a *newly* produced asset; not on the historical result of what an investment has yielded on its original cost if we look back on its record after its life is over.

If there is an increased investment in any given type of capital during any period of time, the marginal efficiency of that type of capital will diminish as the investment in it is increased, partly because the prospective yield will fall as the supply of that type of capital is increased, and partly because, as a rule, pressure on the facilities for producing that type of capital will cause its supply price to increase; the second of these factors being usually the more important in producing equilibrium in the short run, but the longer the period in view the more does the first factor take its place. Thus for each type of capital we can build up a schedule, showing by how much investment in it will have to increase within the period, in order that its marginal efficiency should fall to any given figure. We can then aggregate these schedules for all the different types of capital, so as to provide a schedule relating the rate of aggregate investment to the corresponding marginal efficiency of capital in general which that rate of investment will establish. We shall call this the investment demand-schedule; or, alternatively, the schedule of the marginal efficiency of capital.

Now it is obvious that the actual rate of current investment will be pushed to the point where there is no longer any class of capital-asset of which the marginal efficiency exceeds the current rate of interest. In other words, the rate of investment will be pushed to the point on the investment demand-schedule where the marginal efficiency of capital in general is equal to the market rate of interest.[1]

The same thing can also be expressed as follows. If Q_r is the prospective yield from an asset at time r, and d_r is the present value of £1 deferred r years *at the current rate of interest*, $\Sigma Q_r d_r$ is the demand price of the investment; and investment will be carried to the point where $\Sigma Q_r d_r$ becomes equal to the supply price of the investment as defined above. If, on the other hand, $\Sigma Q_r d_r$ falls short of the supply price, there will be no current investment in the asset in question.

It follows that the inducement to invest depends partly on the investment demand-schedule and partly on the rate of interest. Only at the conclusion of Book IV will it be possible to take a comprehensive view of the factors determining the rate of investment in their actual complexity. I would, however, ask the reader to note at once that neither the knowledge of an asset's prospective yield nor the knowledge of the marginal efficiency of the asset enables us to deduce either the rate of interest or the present value of the asset. We must ascertain the rate of interest from some other source, and only then can we value the asset by 'capitalising' its prospective yield

II

How is the above definition of the marginal efficiency of capital related to common usage? The *Marginal Productivity* or *Yield* or *Efficiency* or *Utility* of Capital are familiar terms which we have all frequently used. But it is not easy by searching the literature of economics to

[1] For the sake of simplicity of statement I have slurred the point that we are dealing with complexes of rates of interest and discount corresponding to the different lengths of time which will elapse before the various prospective returns from the asset are realised. But it is not difficult to re-state the argument so as to cover this point.

find a clear statement of what economists have usually intended by these terms.

There are at least three ambiguities to clear up. There is, to begin with, the ambiguity whether we are concerned with the increment of physical product per unit of time due to the employment of one more physical unit of capital, or with the increment of value due to the employment of one more value unit of capital. The former involves difficulties as to the definition of the physical unit of capital, which I believe to be both insoluble and unnecessary. It is, of course, possible to say that ten labourers will raise more wheat from a given area when they are in a position to make use of certain additional machines; but I know no means of reducing this to an intelligible arithmetical ratio which does not bring in values. Nevertheless many discussions of this subject seem to be mainly concerned with the physical productivity of capital in some sense, though the writers fail to make themselves clear.

Secondly, there is the question whether the marginal efficiency of capital is some absolute quantity or a ratio. The contexts in which it is used and the practice of treating it as being of the same dimension as the rate of interest seem to require that it should be a ratio. Yet it is not usually made clear what the two terms of the ratio are supposed to be.

Finally, there is the distinction, the neglect of which has been the main cause of confusion and misunderstanding, between the increment of value obtainable by using an additional quantity of capital in the *existing* situation, and the series of increments which it is expected to obtain *over the whole life* of the additional capital asset;—i.e. the distinction between Q_1 and the complete series $Q_1, Q_2, \ldots Q_r, \ldots$. This involves the whole question of the place of expectation in economic theory. Most discussions of the marginal efficiency of capital seem to pay no attention to any member of the series except Q_1. Yet this cannot be legitimate except in a static theory, for which all the Q's are equal. The ordinary theory of distribution, where it is assumed that capital is getting *now* its marginal productivity (in some sense or other), is only valid in a stationary state. The aggregate current return to capital has no direct relationship to its marginal efficiency; whilst its current return at the margin of production (i.e. the return to capital which enters into the supply price of output) is its marginal user cost, which also has no close connection with its marginal efficiency.

There is, as I have said above, a remarkable lack of any clear account of the matter. At the same time I believe that the definition which I have given above is fairly close to what Marshall intended to mean by the term. The phrase which Marshall himself uses is 'marginal net efficiency' of a factor of production; or, alternatively, the 'marginal utility of capital'. The following is a summary of the most relevant passage which I can find in his *Principles* (6th ed. pp. 519–520). I have run together some non-consecutive sentences to convey the gist of what he says:

In a certain factory an extra £100 worth of machinery can be applied so as not to involve any other extra expense, and so as to add annually £3 worth to the net output of the factory after allowing for its own wear and tear. If the investors of capital push it into every occupation in which it seems likely to gain a high reward; and if, after this has been done and equilibrium has been found, it still pays and only just pays to employ this machinery, we can infer from this fact that the yearly rate of interest is 3 per cent. But illustrations of this kind merely indicate part of the action of the great causes which govern value. They cannot be made into a theory of interest, any more than into a theory of wages, without reasoning in a circle...Suppose that the rate of interest is 3 per cent. per annum on perfectly good security; and that the hat-making trade absorbs a capital of one million pounds. This implies that the hat-making trade can turn the whole million pounds' worth of capital to so good account that they would pay 3 per cent. per annum net for the use of it rather than go without any of it. There may be machinery which the trade would have refused to dispense with if the rate of interest had been 20 per cent. per annum. If the rate had been 10 per cent., more would have been used; if it had been 6 per cent., still more; if 4 per cent. still more; and finally, the rate being 3 per cent., they use more still. When they have this amount, the marginal utility of the machinery, i.e. the utility of that machinery which it is only just worth their while to employ, is measured by 3 per cent.

It is evident from the above that Marshall was well aware that we are involved in a circular argument if we try to determine along these lines what the rate of interest actually is.[1] In this passage he appears to accept the view set forth above, that the rate of interest determines the point to which new investment will be pushed, given the schedule of the marginal efficiency of capital. If the rate of interest is 3 per cent, this means that no one will pay £100 for a machine unless he hopes thereby to add £3 to his annual net output after allowing for costs and depreciation. But we shall see in chapter 14 that in other passages Marshall was less cautious—though still drawing back when his argument was leading him on to dubious ground.

Although he does not call it the 'marginal efficiency of capital', Professor Irving Fisher has given in his *Theory of Interest* (1930) a definition of what he calls 'the rate of return over cost' which is identical with my definition. 'The rate of return over cost', he writes,[2] 'is that rate which, employed in computing the present worth of all the costs and the present worth of all the returns, will make these two equal.' Professor Fisher explains that the extent of investment in any direction will depend on a comparison between the rate of return over cost and the rate of interest. To induce new investment 'the rate of return over cost

---

[1] But was he not wrong in supposing that the marginal productivity theory of wages is equally circular?

[2] *Op. cit.* p. 168.

must exceed the rate of interest'.[1] 'This new magnitude (or factor) in our study plays the central rôle on the investment opportunity side of interest theory.'[2] Thus Professor Fisher uses his 'rate of return over cost' in the same sense and for precisely the same purpose as I employ 'the marginal efficiency of capital'.

III

The most important confusion concerning the meaning and significance of the marginal efficiency of capital has ensued on the failure to see that it depends on the *prospective* yield of capital, and not merely on its current yield. This can be best illustrated by pointing out the effect on the marginal efficiency of capital of an expectation of changes in the prospective cost of production, whether these changes are expected to come from changes in labour cost, i.e. in the wage-unit, or from inventions and new technique. The output from equipment produced to-day will have to compete, in the course of its life, with the output from equipment produced subsequently, perhaps at a lower labour cost, perhaps by an improved technique, which is content with a lower price for its output and will be increased in quantity until the price of its output has fallen to the lower figure with which it is content. Moreover, the entrepreneur's profit (in terms of money) from equipment, old or new, will be reduced, if all output comes to be produced more cheaply. In so far as such developments are foreseen as probable, or even as possible, the marginal efficiency of capital produced to-day is appropriately diminished.

This is the factor through which the expectation of changes in the value of money influences the volume of current output. The expectation of a fall in the value of money stimulates investment, and hence employment generally, because it raises the schedule of the

---

[1] *Op. cit.* p. 159.   [2] *Op. cit.* p. 155.

marginal efficiency of capital, i.e. the investment demand-schedule; and the expectation of a rise in the value of money is depressing, because it lowers the schedule of the marginal efficiency of capital.

This is the truth which lies behind Professor Irving Fisher's theory of what he originally called 'Appreciation and Interest'—the distinction between the money rate of interest and the real rate of interest where the latter is equal to the former after correction for changes in the value of money. It is difficult to make sense of this theory as stated, because it is not clear whether the change in the value of money is or is not assumed to be foreseen. There is no escape from the dilemma that, if it is not foreseen, there will be no effect on current affairs; whilst, if it is foreseen, the prices of existing goods will be forthwith so adjusted that the advantages of holding money and of holding goods are again equalised, and it will be too late for holders of money to gain or to suffer a change in the rate of interest which will offset the prospective change during the period of the loan in the value of the money lent. For the dilemma is not successfully escaped by Professor Pigou's expedient of supposing that the prospective change in the value of money is foreseen by one set of people but not foreseen by another.

The mistake lies in supposing that it is the rate of interest on which prospective changes in the value of money will directly react, instead of the marginal efficiency of a given stock of capital. The prices of *existing* assets will always adjust themselves to changes in expectation concerning the prospective value of money. The significance of such changes in expectation lies in their effect on the readiness to produce *new* assets through their reaction on the marginal efficiency of capital. The stimulating effect of the expectation of higher prices is due, not to its raising the rate of interest (that would be a paradoxical way of stimulating output—in so far as the rate of interest rises, the stimulating effect is to that extent offset), but to its raising the marginal efficiency of a given stock of capital. *If* the rate of interest were to rise *pari passu* with the marginal efficiency of capital, there would be *no* stimulating effect from the expectation of rising prices. For the stimulus to output depends on the marginal efficiency of a given stock of capital rising *relatively* to the rate of interest. Indeed Professor Fisher's theory could be best re-written in terms of a 'real rate of interest' defined as being the rate of interest which would have to rule, consequently on a change in the state of expectation as to the future value of money, in order that this change should have no effect on current output.[1]

It is worth noting that an expectation of a future fall in the rate of interest will have the effect of *lowering* the schedule of the marginal efficiency of capital; since it means that the output from equipment produced to-day will have to compete during part of its life with the output from equipment which is content with a lower return. This expectation will have no great depressing effect, since the expectations, which are held concerning the complex of rates of interest for various terms which will rule in the future, will be partially reflected in the complex of rates of interest which rule to-day. Nevertheless there may be some depressing effect, since the output from equipment produced to-day, which will emerge towards the end of the life of this equipment, may have to compete with the output of much younger equipment which is content with a lower return because of the lower rate of interest which rules for periods subsequent to the end of the life of equipment produced to-day.

It is important to understand the dependence of the marginal efficiency of a given stock of capital on changes in expectation, because it is chiefly this depend-

---

[1] Cf. Mr Robertson's article on 'Industrial Fluctuations and the Natural Rate of Interest', *Economic Journal*, December 1934.

ence which renders the marginal efficiency of capital subject to the somewhat violent fluctuations which are the explanation of the trade cycle. In chapter 22 below we shall show that the succession of boom and slump can be described and analysed in terms of the fluctuations of the marginal efficiency of capital relatively to the rate of interest.

IV

Two types of risk affect the volume of investment which have not commonly been distinguished, but which it is important to distinguish. The first is the entrepreneur's or borrower's risk and arises out of doubts in his own mind as to the probability of his actually earning the prospective yield for which he hopes. If a man is venturing his own money, this is the only risk which is relevant.

But where a system of borrowing and lending exists, by which I mean the granting of loans with a margin of real or personal security, a second type of risk is relevant which we may call the lender's risk. This may be due either to moral hazard, i.e. voluntary default or other means of escape, possibly lawful, from the fulfilment of the obligation, or to the possible insufficiency of the margin of security, i.e. involuntary default due to the disappointment of expectation. A third source of risk might be added, namely, a possible adverse change in the value of the monetary standard which renders a money-loan to this extent less secure than a real asset; though all or most of this should be already reflected, and therefore absorbed, in the price of durable real assets.

Now the first type of risk is, in a sense, a real social cost, though susceptible to diminution by averaging as well as by an increased accuracy of foresight. The second, however, is a pure addition to the cost of investment which would not exist if the borrower and lender were the same person. Moreover, it involves in part a duplication of a proportion of the entrepreneur's risk, which is added *twice* to the pure rate of interest to give the minimum prospective yield which will induce the investment. For if a venture is a risky one, the borrower will require a wider margin between his expectation of yield and the rate of interest at which he will think it worth his while to borrow; whilst the very same reason will lead the lender to require a wider margin between what he charges and the pure rate of interest in order to induce him to lend (except where the borrower is so strong and wealthy that he is in a position to offer an exceptional margin of security). The hope of a very favourable outcome, which may balance the risk in the mind of the borrower, is not available to solace the lender.

This duplication of allowance for a portion of the risk has not hitherto been emphasised, so far as I am aware; but it may be important in certain circumstances. During a boom the popular estimation of the magnitude of both these risks, both borrower's risk and lender's risk, is apt to become unusually and imprudently low.

V

The schedule of the marginal efficiency of capital is of fundamental importance because it is mainly through this factor (much more than through the rate of interest) that the expectation of the future influences the present. The mistake of regarding the marginal efficiency of capital primarily in terms of the *current* yield of capital equipment, which would be correct only in the static state where there is no changing future to influence the present, has had the result of breaking the theoretical link between to-day and to-morrow. Even the rate of interest is, virtually,[1] a

---

[1] Not completely; for its value partly reflects the *uncertainty* of the future. Moreover, the relation between rates of interest for different terms depends on expectations.

current phenomenon; and if we reduce the marginal efficiency of capital to the same status, we cut ourselves off from taking any direct account of the influence of the future in our analysis of the existing equilibrium.

The fact that the assumptions of the static state often underlie present-day economic theory, imports into it a large element of unreality. But the introduction of the concepts of user cost and of the marginal efficiency of capital, as defined above, will have the effect, I think, of bringing it back to reality, whilst reducing to a minimum the necessary degree of adaptation.

It is by reason of the existence of durable equipment that the economic future is linked to the present. It is, therefore, consonant with, and agreeable to, our broad principles of thought, that the expectation of the future should affect the present through the demand price for durable equipment.

Chapter 12

THE STATE OF LONG-TERM EXPECTATION

I

We have seen in the previous chapter that the scale of investment depends on the relation between the rate of interest and the schedule of the marginal efficiency of capital corresponding to different scales of current investment, whilst the marginal efficiency of capital depends on the relation between the supply price of a capital-asset and its prospective yield. In this chapter we shall consider in more detail some of the factors which determine the prospective yield of an asset.

The considerations upon which expectations of prospective yields are based are partly existing facts which we can assume to be known more or less for certain, and partly future events which can only be forecasted with more or less confidence. Amongst the first may be mentioned the existing stock of various types of capital-assets and of capital-assets in general and the strength of the existing consumers' demand for goods which require for their efficient production a relatively larger assistance from capital. Amongst the latter are future changes in the type and quantity of the stock of capital-assets and in the tastes of the consumer, the strength of effective demand from time to time during the life of the investment under consideration, and the changes in the wage-unit in terms of money which may occur during its life. We may sum up the state of psychological expectation which covers the latter as being the *state of long-term expectation*;—as distinguished from the short-term expectation upon the basis of which a producer estimates what he will get for a product when it is finished if he decides to begin producing it to-day with the existing plant, which we examined in chapter 5.

II

It would be foolish, in forming our expectations, to attach great weight to matters which are very uncertain.[1] It is reasonable, therefore, to be guided to a considerable degree by the facts about which we feel somewhat confident, even though they may be less decisively relevant to the issue than other facts about which our knowledge is vague and scanty. For this reason the facts of the existing situation enter, in a sense disproportionately, into the formation of our long-term expectations; our usual practice being to take the existing situation and to project it into the future, modified only to the extent that we have more or less definite reasons for expecting a change.

The state of long-term expectation, upon which our decisions are based, does not solely depend, therefore, on the most probable forecast we can make. It also depends on the *confidence* with which we make this forecast—on how highly we rate the likelihood of our best forecast turning out quite wrong. If we expect large changes but are very uncertain as to what precise form these changes will take, then our confidence will be weak.

The *state of confidence*, as they term it, is a matter to which practical men always pay the closest and most anxious attention. But economists have not analysed it carefully and have been content, as a rule, to discuss

---

[1] By 'very uncertain' I do not mean the same thing as 'very improbable'. Cf. my *Treatise on Probability* [*JMK*, vol. VIII], chap. 6, on 'The Weight of Arguments'.

it in general terms. In particular it has not been made clear that its relevance to economic problems comes in through its important influence on the schedule of the marginal efficiency of capital. There are not two separate factors affecting the rate of investment, namely, the schedule of the marginal efficiency of capital and the state of confidence. The state of confidence is relevant because it is one of the major factors determining the former, which is the same thing as the investment demand-schedule.

There is, however, not much to be said about the state of confidence *a priori*. Our conclusions must mainly depend upon the actual observation of markets and business psychology. This is the reason why the ensuing digression is on a different level of abstraction from most of this book.

For convenience of exposition we shall assume in the following discussion of the state of confidence that there are no changes in the rate of interest; and we shall write, throughout the following sections, as if changes in the values of investments were solely due to changes in the expectation of their prospective yields and not at all to changes in the rate of interest at which these prospective yields are capitalised. The effect of changes in the rate of interest is, however, easily superimposed on the effect of changes in the state of confidence.

III

The outstanding fact is the extreme precariousness of the basis of knowledge on which our estimates of prospective yield have to be made. Our knowledge of the factors which will govern the yield of an investment some years hence is usually very slight and often negligible. If we speak frankly, we have to admit that our basis of knowledge for estimating the yield ten years hence of a railway, a copper mine, a textile factory, the goodwill of a patent medicine, an Atlantic liner, a building in the City of London amounts to little and sometimes to nothing; or even five years hence. In fact, those who seriously attempt to make any such estimate are often so much in the minority that their behaviour does not govern the market.

In former times, when enterprises were mainly owned by those who undertook them or by their friends and associates, investment depended on a sufficient supply of individuals of sanguine temperament and constructive impulses who embarked on business as a way of life, not really relying on a precise calculation of prospective profit. The affair was partly a lottery, though with the ultimate result largely governed by whether the abilities and character of the managers were above or below the average. Some would fail and some would succeed. But even after the event no one would know whether the average results in terms of the sums invested had exceeded, equalled, or fallen short of the prevailing rate of interest; though, if we exclude the exploitation of natural resources and monopolies, it is probable that the actual average results of investments, even during periods of progress and prosperity, have disappointed the hopes which prompted them. Business men play a mixed game of skill and chance, the average results of which to the players are not known by those who take a hand. If human nature felt no temptation to take a chance, no satisfaction (profit apart) in constructing a factory, a railway, a mine or a farm, there might not be much investment merely as a result of cold calculation.

Decisions to invest in private business of the old-fashioned type were, however, decisions largely irrevocable, not only for the community as a whole, but also for the individual. With the separation between ownership and management which prevails to-day and with the development of organised investment markets, a new factor of great importance has entered in, which sometimes facilitates investment but sometimes adds greatly to the instability of the system. In the absence of security markets, there is no object in frequently attempting to revalue an investment to which we are committed. But the Stock Exchange revalues many investments every day and the revaluations give a frequent opportunity to the individual (though not to the community as a whole) to revise his commitments. It is as though a farmer, having tapped his barometer after breakfast, could decide to remove his capital from the farming business between 10 and 11 in the morning and reconsider whether he should return to it later in the week. But the daily revaluations of the Stock Exchange, though they are primarily made to facilitate transfers of old investments between one individual and another, inevitably exert a decisive influence on the rate of current investment. For there is no sense in building up a new enterprise at a cost greater than that at which a similar existing enterprise can be purchased; whilst there is an inducement to spend on a new project what may seem an extravagant sum, if it can be floated off on the Stock Exchange at an immediate profit.[1] Thus certain classes of investment are governed by the average expectation of those who deal on the Stock Exchange as revealed in the price of shares, rather than by the genuine expectations of the professional entrepreneur.[2] How then are these highly significant daily, even hourly, revaluations of existing investments carried out in practice?

[1]   In my *Treatise on Money* (vol. ii. p. 195) [*JMK*, vol. vi, p. 174] I pointed out that when a company's shares are quoted very high so that it can raise more capital by issuing more shares on favourable terms, this has the same effect as if it could borrow at a low rate of interest. I should now describe this by saying that a high quotation for existing equities involves an increase in the marginal efficiency of the corresponding type of capital and therefore has the same effect (since investment depends on a comparison between the marginal efficiency of capital and the rate of interest) as a fall in the rate of interest.

[2]   This does not apply, of course, to classes of enterprise which are not readily marketable or to which no negotiable instrument closely corresponds. The categories falling within this exception were formerly extensive. But measured as a proportion of the total value of new investment they are rapidly declining in importance.

IV

In practice we have tacitly agreed, as a rule, to fall back on what is, in truth, a *convention*. The essence of this convention—though it does not, of course, work out quite so simply—lies in assuming that the existing state of affairs will continue indefinitely, except in so far as we have specific reasons to expect a change. This does not mean that we really believe that the existing state of affairs will continue indefinitely. We know from extensive experience that this is most unlikely. The actual results of an investment over a long term of years very seldom agree with the initial expectation. Nor can we rationalise our behaviour by arguing that to a man in a state of ignorance errors in either direction are equally probable, so that there remains a mean actuarial expectation based on equi-probabilities. For it can easily be shown that the assumption of arithmetically equal probabilities based on a state of ignorance leads to absurdities. We are assuming, in effect, that the existing market valuation, however arrived at, is uniquely *correct* in relation to our existing knowledge of the facts which will influence the yield of the investment, and that it will only change in proportion to changes in this knowledge; though, philosophically speaking, it cannot be uniquely correct, since our existing knowledge does not provide a sufficient basis for a calculated mathematical expectation. In point of fact, all sorts of considerations enter into the market valuation which are in no way relevant to the prospective yield.

Nevertheless the above conventional method of calculation will be compatible with a considerable measure of continuity and stability in our affairs, *so long as we can rely on the maintenance of the convention*.

For if there exist organised investment markets and if we can rely on the maintenance of the convention, an investor can legitimately encourage himself with the idea that the only risk he runs is that of a genuine change in the news *over the near future*, as to the likelihood of which he can attempt to form his own judgment, and which is unlikely to be very large. For, assuming that the convention holds good, it is only these changes which can affect the value of his investment, and he need not lose his sleep merely because he has not any notion what his investment will be worth ten years hence. Thus investment becomes reasonably 'safe' for the individual investor over short periods, and hence over a succession of short periods however many, if he can fairly rely on there being no breakdown in the convention and on his therefore having an opportunity to revise his judgment and change his investment, before there has been time for much to happen. Investments which are 'fixed' for the community are thus made 'liquid' for the individual.

It has been, I am sure, on the basis of some such procedure as this that our leading investment markets have been developed. But it is not surprising that a convention, in an absolute view of things so arbitrary, should have its weak points. It is its precariousness which creates no small part of our contemporary problem of securing sufficient investment.

V

Some of the factors which accentuate this precariousness may be briefly mentioned.

(1) As a result of the gradual increase in the proportion of the equity in the community's aggregate capital investment which is owned by persons who do not manage and have no special knowledge of the circumstances, either actual or prospective, of the business in question, the element of real knowledge in the valuation of investments by whose who own them or contemplate purchasing them has seriously declined.

(2) Day-to-day fluctuations in the profits of existing

investments, which are obviously of an ephemeral and non-significant character, tend to have an altogether excessive, and even an absurd, influence on the market. It is said, for example, that the shares of American companies which manufacture ice tend to sell at a higher price in summer when their profits are seasonally high than in winter when no one wants ice. The recurrence of a bank-holiday may raise the market valuation of the British railway system by several million pounds.

(3) A conventional valuation which is established as the outcome of the mass psychology of a large number of ignorant individuals is liable to change violently as the result of a sudden fluctuation of opinion due to factors which do not really make much difference to the prospective yield; since there will be no strong roots of conviction to hold it steady. In abnormal times in particular, when the hypothesis of an indefinite continuance of the existing state of affairs is less plausible than usual even though there are no express grounds to anticipate a definite change, the market will be subject to waves of optimistic and pessimistic sentiment, which are unreasoning and yet in a sense legitimate where no solid basis exists for a reasonable calculation.

(4) But there is one feature in particular which deserves our attention. It might have been supposed that competition between expert professionals, possessing judgment and knowledge beyond that of the average private investor, would correct the vagaries of the ignorant individual left to himself. It happens, however, that the energies and skill of the professional investor and speculator are mainly occupied otherwise. For most of these persons are, in fact, largely concerned, not with making superior long-term forecasts of the probable yield of an investment over its whole life, but with foreseeing changes in the conventional basis of valuation a short time ahead of the general public. They are concerned, not with what an investment is

really worth to a man who buys it 'for keeps', but with what the market will value it at, under the influence of mass psychology, three months or a year hence. Moreover, this behaviour is not the outcome of a wrong-headed propensity. It is an inevitable result of an investment market organised along the lines described. For it is not sensible to pay 25 for an investment of which you believe the prospective yield to justify a value of 30, if you also believe that the market will value it at 20 three months hence.

Thus the professional investor is forced to concern himself with the anticipation of impending changes, in the news or in the atmosphere, of the kind by which experience shows that the mass psychology of the market is most influenced. This is the inevitable result of investment markets organised with a view to so-called 'liquidity'. Of the maxims of orthodox finance none, surely, is more anti-social than the fetish of liquidity, the doctrine that it is a positive virtue on the part of investment institutions to concentrate their resources upon the holding of 'liquid' securities. It forgets that there is no such thing as liquidity of investment for the community as a whole. The social object of skilled investment should be to defeat the dark forces of time and ignorance which envelop our future. The actual, private object of the most skilled invest-ment to-day is 'to beat the gun', as the Americans so well express it, to outwit the crowd, and to pass the bad, or depreciating, half-crown to the other fellow.

This battle of wits to anticipate the basis of con-ventional valuation a few months hence, rather than the prospective yield of an investment over a long term of years, does not even require gulls amongst the public to feed the maws of the professional;—it can be played by professionals amongst themselves. Nor is it necessary that anyone should keep his simple faith in the con-ventional basis of valuation having any genuine long-term validity. For it is, so to speak, a game of Snap, of Old Maid, of Musical Chairs—a pastime in which he is victor who says *Snap* neither too soon nor too late, who passed the Old Maid to his neighbour before the game is over, who secures a chair for himself when the music stops. These games can be played with zest and enjoyment, though all the players know that it is the Old Maid which is circulating, or that when the music stops some of the players will find themselves unseated.

Or, to change the metaphor slightly, professional investment may be likened to those newspaper competitions in which the competitors have to pick out the six prettiest faces from a hundred photographs, the prize being awarded to the competitor whose choice most nearly corresponds to the average preferences of the competitors as a whole; so that each competitor has to pick, not those faces which he himself finds prettiest, but those which he thinks likeliest to catch the fancy of the other competitors, all of whom are looking at the problem from the same point of view. It is not a case of choosing those which, to the best of one's judgment, are really the prettiest, nor even those which average opinion genuinely thinks the prettiest. We have reached the third degree where we devote our intelligences to anticipating what average opinion expects the average opinion to be. And there are some, I believe, who practise the fourth, fifth and higher degrees.

If the reader interjects that there must surely be large profits to be gained from the other players in the long run by a skilled individual who, unperturbed by the prevailing pastime, continues to purchase investments on the best genuine long-term expectations he can frame, he must be answered, first of all, that there are, indeed, such serious-minded individuals and that it makes a vast difference to an investment market whether or not they predominate in their influence over the game-players. But we must also add that there are

several factors which jeopardise the predominance of
such individuals in modern investment markets. Investment based on genuine long-term expectation is so
difficult to-day as to be scarcely practicable. He who
attempts it must surely lead much more laborious days
and run greater risks than he who tries to guess better
than the crowd how the crowd will behave; and, given
equal intelligence, he may make more disastrous mistakes. There is no clear evidence from experience that
the investment policy which is socially advantageous
coincides with that which is most profitable. It needs
more intelligence to defeat the forces of time and
our ignorance of the future than to beat the gun.
Moreover, life is not long enough;—human nature
desires quick results, there is a peculiar zest in making
money quickly, and remoter gains are discounted by the
average man at a very high rate. The game of professional investment is intolerably boring and over-
exacting to anyone who is entirely exempt from the
gambling instinct; whilst he who has it must pay to
this propensity the appropriate toll. Furthermore, an
investor who proposes to ignore near-term market
fluctuations needs greater resources for safety and must
not operate on so large a scale, if at all, with borrowed
money—a further reason for the higher return from the
pastime to a given stock of intelligence and resources.
Finally it is the long-term investor, he who most
promotes the public interest, who will in practice come
in for most criticism, wherever investment funds are
managed by committees or boards or banks.[1] For it is
in the essence of his behaviour that he should be
eccentric, unconventional and rash in the eyes of
average opinion. If he is successful, that will only
confirm the general belief in his rashness; and if

[1] The practice, usually considered prudent, by which an investment trust
or an insurance office frequently calculates not only the income from its
investment portfolio but also its capital valuation in the market, may also
tend to direct too much attention to short-term fluctuations in the latter.

in the short run he is unsuccessful, which is very likely, he will not receive much mercy. Worldly wisdom teaches that it is better for reputation to fail conventionally than to succeed unconventionally.

(5) So far we have had chiefly in mind the state of confidence of the speculator or speculative investor himself and may have seemed to be tacitly assuming that, if he himself is satisfied with the prospects, he has unlimited command over money at the market rate of interest. This is, of course, not the case. Thus we must also take account of the other facet of the state of confidence, namely, the confidence of the lending institutions towards those who seek to borrow from them, sometimes described as the state of credit. A collapse in the price of equities, which has had disastrous reactions on the marginal efficiency of capital, may have been due to the weakening either of speculative confidence or of the state of credit. But whereas the weakening of either is enough to cause a collapse, recovery requires the revival of *both*. For whilst the weakening of credit is sufficient to bring about a collapse, its strengthening, though a necessary condition of recovery, is not a sufficient condition.

VI

These considerations should not lie beyond the purview of the economist. But they must be relegated to their right perspective. If I may be allowed to appropriate the term *speculation* for the activity of forecasting the psychology of the market, and the term *enterprise* for the activity of forecasting the prospective yield of assets over their whole life, it is by no means always the case that speculation predominates over enterprise. As the organisation of investment markets improves, the risk of the predominance of speculation does, however, increase. In one of the greatest investment markets in the world, namely, New York, the influence of speculation (in the above sense) is enormous. Even outside the field of finance, Americans are apt to be unduly interested in discovering what average opinion believes average opinion to be; and this national weakness finds its nemesis in the stock market. It is rare, one is told, for an American to invest, as many Englishmen still do, 'for income'; and he will not readily purchase an investment except in the hope of capital appreciation. This is only another way of saying that, when he purchases an investment, the American is attaching his hopes, not so much to its prospective yield, as to a favourable change in the conventional basis of valuation, i.e. that he is, in the above sense, a speculator. Speculators may do no harm as bubbles on a steady stream of enterprise. But the position is serious when enterprise becomes the bubble on a whirl-pool of speculation. When the capital development of a country becomes a by-product of the activities of a casino, the job is likely to be ill-done. The measure of success attained by Wall Street, regarded as an institution of which the proper social purpose is to direct new investment into the most profitable channels in terms of future yield, cannot be claimed as one of the outstanding triumphs of *laissez-faire* capitalism—which is not surprising, if I am right in thinking that the best brains of Wall Street have been in fact directed towards a different object.

These tendencies are a scarcely avoidable outcome of our having successfully organised 'liquid' investment markets. It is usually agreed that casinos should, in the public interest, be inaccessible and expensive. And perhaps the same is true of stock exchanges. That the sins of the London Stock Exchange are less than those of Wall Street may be due, not so much to differences in national character, as to the fact that to the average Englishman Throgmorton Street is, compared with Wall Street to the average American, inaccessible and very expensive. The jobber's 'turn', the high brokerage charges and the heavy transfer tax payable to the Exchequer, which attend dealings on the London Stock Exchange, sufficiently diminish the liquidity of the market (although the practice of fortnightly accounts operates the other way) to rule out a large proportion of the transaction characteristic of Wall Street.[1] The introduction of a substantial government transfer tax on all transactions might prove the most serviceable reform available, with a view to mitigating the predominance of speculation over enterprise in the United States.

The spectacle of modern investment markets has sometimes moved me towards the conclusion that to make the purchase of an investment permanent and indissoluble, like marriage, except by reason of death or other grave cause, might be a useful remedy for our contemporary evils. For this would force the investor to direct his mind to the long-term prospects and to those only. But a little consideration of this expedient brings us up against a dilemma, and shows us how the liquidity of investment markets often facilitates, though it sometimes impedes, the course of new investment. For the fact that each individual investor flatters himself that his commitment is 'liquid' (though this cannot be true for all investors collectively) calms his nerves and makes him much more willing to run a risk. If individual purchases of investments were rendered illiquid, this might seriously impede new investment, so long as *alternative ways* in which to hold his savings are available to the individual. This is the dilemma. So long as it is open to the individual to employ his wealth in hoarding or lending *money*, the alternative of purchasing actual capital assets cannot be rendered sufficiently attractive (especially to the man who does

[1] It is said that, when Wall Street is active, at least a half of the purchases or sales of investments are entered upon with an intention on the part of the speculator to reverse them *the same day*. This is often true of the commodity exchanges also.

not manage the capital assets and knows very little about them), except by organising markets wherein these assets can be easily realised for money.

The only radical cure for the crises of confidence which afflict the economic life of the modern world would be to allow the individual no choice between consuming his income and ordering the production of the specific capital-asset which, even though it be on precarious evidence, impresses him as the most promising investment available to him. It might be that, at times when he was more than usually assailed by doubts concerning the future, he would turn in his perplexity towards more consumption and less new investment. But that would avoid the disastrous, cumulative and far-reaching repercussions of its being open to him, when thus assailed by doubts, to spend his income neither on the one nor on the other.

Those who have emphasised the social dangers of the hoarding of money have, of course, had something similar to the above in mind. But they have overlooked the possibility that the phenomenon can occur without any change, or at least any commensurate change, in the hoarding of money

VII

Even apart from the instability due to speculation, there is the instability due to the characteristic of human nature that a large proportion of our positive activities depend on spontaneous optimism rather than on a mathematical expectation, whether moral or hedonistic or economic. Most, probably, of our decisions to do something positive, the full consequences of which will be drawn out over many days to come, can only be taken as a result of animal spirits—of a spontaneous urge to action rather than inaction, and not as the outcome of a weighted average of quantitative benefits multiplied by quantitative probabilities. Enterprise only pretends to itself to be mainly actuated by the statements in its own prospectus, however candid and sincere. Only a little more than an expedition to the South Pole, is it based on an exact calculation of benefits to come. Thus if the animal spirits are dimmed and the spontaneous optimism falters, leaving us to depend on nothing but a mathematical expectation, enterprise will fade and die;—though fears of loss may have a basis no more reasonable than hopes of profit had before.

It is safe to say that enterprise which depends on hopes stretching into the future benefits the community as a whole. But individual initiative will only be adequate when reasonable calculation is supplemented and supported by animal spirits, so that the thought of ultimate loss which often overtakes pioneers, as experience undoubtedly tells us and them, is put aside as a healthy man puts aside the expectation of death.

This means, unfortunately, not only that slumps and depressions are exaggerated in degree, but that economic prosperity is excessively dependent on a political and social atmosphere which is congenial to the average business man. If the fear of a Labour Government or a New Deal depresses enterprise, this need not be the result either of a reasonable calculation or of a plot with political intent;—it is the mere consequence of upsetting the delicate balance of spontaneous optimism. In estimating the prospects of investment, we must have regard, therefore, to the nerves and hysteria and even the digestions and reactions to the weather of those upon whose spontaneous activity it largely depends.

We should not conclude from this that everything depends on waves of irrational psychology. On the contrary, the state of long-term expectation is often steady, and, even when it is not, the other factors exert their compensating effects. We are merely reminding ourselves that human decisions affecting the future, whether personal or political or economic, cannot depend on strict mathematical expectation, since the basis for making such calculations does not exist; and that it is our innate urge to activity which makes the wheels go round, our rational selves choosing between the alternatives as best we are able, calculating where we can, but often falling back for our motive on whim or sentiment or chance.

VIII

There are, moreover, certain important factors which somewhat mitigate in practice the effects of our ignorance of the future. Owing to the operation of compound interest combined with the likelihood of obsolescence with the passage of time, there are many individual investments of which the prospective yield is legitimately dominated by the returns of the comparatively near future. In the case of the most important class of very long-term investments, namely buildings, the risk can be frequently transferred from the investor to the occupier, or at least shared between them, by means of long-term contracts, the risk being outweighed in the mind of the occupier by the advantages of continuity and security of tenure. In the case of another important class of long-term investments, namely public utilities, a substantial proportion of the prospective yield is practically guaranteed by monopoly privileges coupled with the right to charge such rates as will provide a certain stipulated margin. Finally there is a growing class of investments entered upon by, or at the risk of, public authorities, which are frankly influenced in making the investment by a general presumption of there being prospective social advantages from the investment, whatever its commercial yield may prove to be within a wide range, and without seeking to be satisfied that the mathematical expectation of the yield is at least equal to the current rate of interest,—though the rate which the public

163

authority has to pay may still play a decisive part in determining the scale of investment operations which it can afford.

Thus after giving full weight to the importance of the influence of short-period changes in the state of long-term expectation as distinct from changes in the rate of interest, we are still entitled to return to the latter as exercising, at any rate, in normal circumstances, a great, though not a decisive, influence on the rate of investment. Only experience, however, can show how far management of the rate of interest is capable of continuously stimulating the appropriate volume of investment.

For my own part I am now somewhat sceptical of the success of a merely monetary policy directed towards influencing the rate of interest. I expect to see the State, which is in a position to calculate the marginal efficiency of capital-goods on long views and on the basis of the general social advantage, taking an ever greater responsibility for directly organising investment; since it seems likely that the fluctuations in the market estimation of the marginal efficiency of different types of capital, calculated on the principles I have described above, will be too great to be offset by any practicable changes in the rate of interest.

Chapter 13

THE GENERAL THEORY OF THE RATE OF INTEREST

I

We have shown in chapter 11 that, whilst there are forces causing the rate of investment to rise or fall so as to keep the marginal efficiency of capital equal to the rate of interest, yet the marginal efficiency of capital is, in itself, a different thing from the ruling rate of interest. The schedule of the marginal efficiency of capital may be said to govern the terms on which loanable funds are demanded for the purpose of new investment; whilst the rate of interest governs the terms on which funds are being currently supplied. To complete our theory, therefore, we need to know what determines the rate of interest.

In chapter 14 and its Appendix we shall consider the answers to this question which have been given hitherto. Broadly speaking, we shall find that they make the rate of interest to depend on the interaction of the schedule of the marginal efficiency of capital with the psychological propensity to save. But the notion that the rate of interest is the balancing factor which brings the demand for saving in the shape of new investment forthcoming at a given rate of interest into equality with the supply of saving which results at that rate of interest from the community's psychological propensity to save, breaks down as soon as we perceive that it is impossible to deduce the rate of interest merely from a knowledge of these two factors.

What, then, is our own answer to this question?

II

The psychological time-preferences of an individual require two distinct sets of decisions to carry them out completely. The first is concerned with that aspect of time-preference which I have called the *propensity to consume*, which, operating under the influence of the various motives set forth in Book III, determines for each individual how much of his income he will consume and how much he will reserve in *some* form of command over future consumption.

But this decision having been made, there is a further decision which awaits him, namely, in *what form* he will hold the command over future consumption which he has reserved, whether out of his current income or from previous savings. Does he want to hold it in the form of immediate, liquid command (i.e. in money or its equivalent)? Or is he prepared to part with immediate command for a specified or indefinite period, leaving it to future market conditions to determine on what terms he can, if necessary, convert deferred command over specific goods into immediate command over goods in general? In other words, what is the degree of his *liquidity-preference*—where an individual's liquidity-preference is given by a schedule of the amounts of his resources, valued in terms of money or of wage-units, which he will wish to retain in the form of money in different sets of circumstances?

We shall find that the mistake in the accepted theories of the rate of interest lies in their attempting to derive the rate of interest from the first of these two constituents of psychological time-preference to the neglect of the second; and it is this neglect which we must endeavour to repair.

It should be obvious that the rate of interest cannot

be a return to saving or waiting as such. For if a man hoards his savings in cash, he earns no interest, though he saves just as much as before. On the contrary, the mere definition of the rate of interest tells us in so many words that the rate of interest is the reward for parting with liquidity for a specified period. For the rate of interest is, in itself, nothing more than the inverse proportion between a sum of money and what can be obtained for parting with control over the money in exchange for a debt[1] for a stated period of time.[2]

Thus the rate of interest at any time, being the reward for parting with liquidity, is a measure of the unwillingness of those who possess money to part with their liquid control over it. The rate of interest is not the 'price' which brings into equilibrium the demand for resources to invest with the readiness to abstain from present consumption. It is the 'price' which equilibrates the desire to hold wealth in the form of cash with the available quantity of cash;—which implies that if the rate of interest were lower, i.e. if the reward for parting with cash were diminished, the aggregate amount of cash which the public would wish to hold would exceed the available supply, and that if the rate of interest were raised, there would be a surplus of cash which no one would be willing to hold. If this explanation is correct, the quantity of money is the

[1] Without disturbance to this definition, we can draw the line between 'money' and 'debts' at whatever point is most convenient for handling a particular problem. For example, we can treat as *money* any command over general purchasing power which the owner has not parted with for a period in excess of three months, and as *debt* what cannot be recovered for a longer period than this; or we can substitute for 'three months' one month or three days or three hours or any other period; or we can exclude from *money* whatever is not legal tender on the spot. It is often convenient in practice to include in *money* time-deposits with banks and, occasionally, even such instruments as (e.g.) treasury bills. As a rule, I shall, as in my *Treatise on Money*, assume that money is co-extensive with bank deposits.

[2] In general discussion, as distinct from specific problems where the period of the debt is expressly specified, it is convenient to mean by the rate of interest the complex of the various rates of interest current for different periods of time, i.e. for debts of different maturities.

other factor, which, in conjunction with liquidity-pre-ference, determines the actual rate of interest in given circumstances. Liquidity-preference is a potentiality or functional tendency, which fixes the quantity of money which the public will hold when the rate of interest is given; so that if r is the rate of interest, M the quantity of money and L the function of liquidity-preference, we have $M = L(r)$. This is where, and how, the quantity of money enters into the economic scheme.

At this point, however, let us turn back and con-sider why such a thing as liquidity-preference exists. In this connection we can usefully employ the ancient distinction between the use of money for the transaction of current business and its use as a store of wealth. As regards the first of these two uses, it is obvious that up to a point it is worth while to sacrifice a certain amount of interest for the convenience of liquidity. But, given that the rate of interest is never negative, why should anyone prefer to hold his wealth in a form which yields little or no interest to holding it in a form which yields interest (assuming, of course, at this stage, that the risk of default is the same in respect of a bank balance as of a bond)? A full explanation is complex and must wait for chapter 15. There is, however, a necessary condition failing which the existence of a liquidity-preference for money as a means of holding wealth could not exist.

This necessary condition is the existence of *un-certainty* as to the future of the rate of interest, i.e. as to the complex of rates of interest for varying maturities which will rule at future dates. For if the rates of interest ruling at all future times could be foreseen with certainty, all future rates of interest could be inferred from the *present* rates of interest for debts of different maturities, which would be adjusted to the knowledge of the future rates. For example, if $_{1}d_{r}$ is the value in the present year 1 of £1 deferred r years and it is known that $_nd_r$ will be the value in the year n of £1 deferred r years from that date, we have

$$_nd_r = \frac{_1d_{n+r}}{_1d_n};$$

whence it follows that the rate at which any debt can be turned into cash n years hence is given by two out of the complex of current rates of interest. If the current rate of interest is positive for debts of every maturity, it must always be more advantageous to purchase a debt than to hold cash as a store of wealth.

If, on the contrary, the future rate of interest is uncertain we cannot safely infer that $_nd_r$ will prove to be equal to $\frac{_1d_{n+r}}{_1d_n}$ when the time comes. Thus if a need for liquid cash may conceivably arise before the expiry of n years, there is a risk of a loss being incurred in purchasing a long-term debt and subsequently turning it into cash, as compared with holding cash. The actuarial profit or mathematical expectation of gain calculated in accordance with the existing probabilities—if it can be so calculated, which is doubtful —must be sufficient to compensate for the risk of disappointment.

There is, moreover, a further ground for liquidity-preference which results from the existence of uncertainty as to the future of the rate of interest, provided that there is an organised market for dealing in debts. For different people will estimate the prospects differently and anyone who differs from the predominant opinion as expressed in market quotations may have a good reason for keeping liquid resources in order to profit, if he is right, from its turning out in due course that the $_1d_r$'s were in a mistaken relationship to one another.[1]

This is closely analogous to what we have already

---

[1] This is the same point as I discussed in my *Treatise on Money* under the designation of the two views and the 'bull-bear' position.

discussed at some length in connection with the marginal efficiency of capital. Just as we found that the marginal efficiency of capital is fixed, not by the 'best' opinion, but by the market valuation as determined by mass psychology, so also expectations as to the future of the rate of interest as fixed by mass psychology have their reactions on liquidity-preference;—but with this addition that the individual, who believes that future rates of interest will be above the rates assumed by the market, has a reason for keeping actual liquid cash,[1] whilst the individual who differs from the market in the other direction will have a motive for borrowing money for short periods in order to purchase debts of longer term. The market price will be fixed at the point at which the sales of the 'bears' and the purchases of the 'bulls' are balanced.

The three divisions of liquidity-preference which we have distinguished above may be defined as depending on (i) the transactions-motive, i.e. the need of cash for the current transaction of personal and business exchanges; (ii) the precautionary-motive, i.e. the desire for security as to the future cash equivalent of a certain proportion of total resources; and (iii) the speculative-motive, i.e. the object of securing profit from knowing better than the market what the future will bring forth. As when we were discussing the marginal efficiency of capital, the question of the desirability of having a highly organised market for dealing with debts presents us with a dilemma. For, in the absence of an organised market, liquidity-preference due to the precautionary-motive would be greatly increased; whereas the existence of an organised market gives an

---

[1] It might be thought that, in the same way, an individual, who believed that the prospective yield of investments will be below what the market is expecting, will have a sufficient reason for holding liquid cash. But this is not the case. He has a sufficient reason for holding cash or debts in preference to equities; but the purchase of debts will be a preferable alternative to holding cash, unless he also believes that the future rate of interest will prove to be higher than the market is supposing.

opportunity for wide fluctuations in liquidity-preference due to the speculative-motive.

It may illustrate the argument to point out that, if the liquidity-preferences due to the transactions-motive and the precautionary-motive are assumed to absorb a quantity of cash which is not very sensitive to changes in the rate of interest as such and apart from its reactions on the level of income, so that the total quantity of money, less this quantity, is available for satisfying liquidity-preferences due to the speculative-motive, the rate of interest and the price of bonds have to be fixed at the level at which the desire on the part of certain individuals to hold cash (because at that level they feel 'bearish' of the future of bonds) is exactly equal to the amount of cash available for the speculative-motive. Thus each increase in the quantity of money must raise the price of bonds sufficiently to exceed the expectations of some 'bull' and so influence him to sell his bond for cash and join the 'bear' brigade. If, however, there is a negligible demand for cash from the speculative-motive except for a short transitional interval, an increase in the quantity of money will have to lower the rate of interest almost forthwith, in whatever degree is necessary to raise employment and the wage-unit sufficiently to cause the additional cash to be absorbed by the transactions-motive and the precautionary-motive.

As a rule, we can suppose that the schedule of liquidity-preference relating the quantity of money to the rate of interest is given by a smooth curve which shows the rate of interest falling as the quantity of money is increased. For there are several different causes all leading towards this result.

In the first place, as the rate of interest falls, it is likely, *cet. par.*, that more money will be absorbed by liquidity-preferences due to the transactions-motive. For if the fall in the rate of interest increases the national income, the amount of money which it is convenient to

keep for transactions will be increased more or less proportionately to the increase in income; whilst, at the
same time, the cost of the convenience of plenty of ready
cash in terms of loss of interest will be diminished.
Unless we measure liquidity-preference in terms of
wage-units rather than of money (which is convenient
in some contexts), similar results follow if the increased
employment ensuing on a fall in the rate of interest
leads to an increase of wages, i.e. to an increase in the
money value of the wage-unit. In the second place,
every fall in the rate of interest may, as we have just
seen, increase the quantity of cash which certain individuals will wish to hold because their views as to the
future of the rate of interest differ from the market
views.

Nevertheless, circumstances can develop in which
even a large increase in the quantity of money may
exert a comparatively small influence on the rate of
interest. For a large increase in the quantity of
money may cause so much uncertainty about the future
that liquidity-preferences due to the precautionary-
motive may be strengthened; whilst opinion about
the future of the rate of interest may be so un-
animous that a small change in present rates may
cause a mass movement into cash. It is interesting
that the stability of the system and its sensitiveness to
changes in the quantity of money should be so dependent
on the existence of a *variety* of opinion about what is
uncertain. Best of all that we should know the future.
But if not, then, if we are to control the activity of the
economic system by changing the quantity of money,
it is important that opinions should differ. Thus
this method of control is more precarious in the
United States, where everyone tends to hold the same
opinion at the same time, than in England where differences of opinion are more usual.

III

We have now introduced money into our causal nexus for the first time, and we are able to catch a first glimpse of the way in which changes in the quantity of money work their way into the economic system. If, however, we are tempted to assert that money is the drink which stimulates the system to activity, we must remind ourselves that there may be several slips between the cup and the lip. For whilst an increase in the quantity of money may be expected, *cet. par.*, to reduce the rate of interest, this will not happen if the liquidity-preferences of the public are increasing more than the quantity of money; and whilst a decline in the rate of interest may be expected, *cet. par.*, to increase the volume of investment, this will not happen if the schedule of the marginal efficiency of capital is falling more rapidly than the rate of interest; and whilst an increase in the volume of investment may be expected, *cet. par.*, to increase employment, this may not happen if the propensity to consume is falling off. Finally, if employment increases, prices will rise in a degree partly governed by the shapes of the physical supply functions, and partly by the liability of the wage-unit to rise in terms of money. And when output has increased and prices have risen, the effect of this on liquidity-preference will be to increase the quantity of money necessary to maintain a given rate of interest.

IV

Whilst liquidity-preference due to the speculative-motive corresponds to what in my *Treatise on Money* I called 'the state of bearishness', it is by no means the same thing. For 'bearishness' is there defined as the functional relationship, not between the rate of interest (or price of debts) and the quantity of money, but between the price of assets and debts, taken together, and the quantity of money. This treatment, however, involved a confusion between results due to a change in the rate of interest and those due to a change in the schedule of the marginal efficiency of capital, which I hope I have here avoided.

V

The concept of *hoarding* may be regarded as a first approximation to the concept of *liquidity-preference*. Indeed if we were to substitute 'propensity to hoard' for 'hoarding', it would come to substantially the same thing. But if we mean by 'hoarding' an actual increase in cash-holding, it is an incomplete idea—and seriously misleading if it causes us to think of 'hoarding' and 'not-hoarding' as simple alternatives. For the decision to hoard is not taken absolutely or without regard to the advantages offered for parting with liquidity;—it results from a balancing of advantages, and we have, therefore, to know what lies in the other scale. Moreover it is impossible for the actual amount of hoarding to change as a result of decisions on the part of the public, so long as we mean by 'hoarding' the actual holding of cash. For the amount of hoarding must be equal to the quantity of money (or—on some definitions—to the quantity of money *minus* what is required to satisfy the transactions-motive); and the quantity of money is not determined by the public. All that the propensity of the public towards hoarding can achieve is to determine the rate of interest at which the aggregate desire to hoard becomes equal to the available cash. The habit of overlooking the relation of the rate of interest to hoarding may be a part of the explanation why interest has been usually regarded as the reward of not-spending, whereas in fact it is the reward of not-hoarding.

Chapter 14

THE CLASSICAL THEORY OF
THE RATE OF INTEREST

I

What is the classical theory of the rate of interest? It is something upon which we have all been brought up and which we have accepted without much reserve until recently. Yet I find it difficult to state it precisely or to discover an explicit account of it in the leading treatises of the modern classical school.[1]

It is fairly clear, however, that this tradition has regarded the rate of interest as the factor which brings the demand for investment and the willingness to save into equilibrium with one another. Investment represents the demand for investible resources and saving represents the supply, whilst the rate of interest is the 'price' of investible resources at which the two are equated. Just as the price of a commodity is necessarily fixed at that point where the demand for it is equal to the supply, so the rate of interest necessarily comes to rest under the play of market forces at the point where the amount of investment at that rate of interest is equal to the amount of saving at that rate.

The above is not to be found in Marshall's *Principles* in so many words. Yet his theory seems to be this, and it is what I myself was brought up on and what I taught for many years to others. Take, for example, the following passage from his *Principles*: 'Interest, being the price paid for the use of capital in any

[1] See the Appendix to this chapter for an abstract of what I have been able to find.

market, tends towards an equilibrium level such that the aggregate demand for capital in that market, at that rate of interest, is equal to the aggregate stock forthcoming at that rate'.[1] Or again in Professor Cassel's *Nature and Necessity of Interest* it is explained that investment constitutes the 'demand for waiting' and saving the 'supply of waiting', whilst interest is a 'price' which serves, it is implied, to equate the two, though here again I have not found actual words to quote. Chapter vi of Professor Carver's *Distribution of Wealth* clearly envisages interest as the factor which brings into equilibrium the marginal disutility of waiting with the marginal productivity of capital.[2] Sir Alfred Flux (*Economic Principles*, p. 95) writes: 'If there is justice in the contentions of our general discussion, it must be admitted that an automatic adjustment takes place between saving and the opportunities for employing capital profitably...Saving will not have exceeded its possibilities of usefulness...so long as the rate of net interest is in excess of zero.' Professor Taussig (*Principles*, vol. ii. p. 29) draws a supply curve of saving and a demand curve representing 'the diminishing productiveness of the several instalments of capital', having previously stated (p. 20) that 'the rate of interest settles at a point where the marginal productivity of capital suffices to bring out the marginal instalment of saving'.[3] Walras, in

---

[1] Cf. p. 186 below for a further discussion of this passage.

[2] Prof. Carver's discussion of interest is difficult to follow (1) through his inconsistency as to whether he means by 'marginal productivity of capital' quantity of marginal product or value of marginal product, and (2) through his making no attempt to define quantity of capital.

[3] In a very recent discussion of these problems ('Capital, Time and the Interest Rate', by Prof. F. H. Knight, *Economica*, August 1934), a discussion which contains many interesting and profound observations on the nature of capital, and confirms the soundness of the Marshallian tradition as to the uselessness of the Böhm–Bawerkian analysis, the theory of interest is given precisely in the traditional, classical mould. Equilibrium in the field of capital production means, according to Prof. Knight, 'such a rate of interest that savings flow into the market at precisely the same time-rate or speed as they flow into investment producing the same net rate of return as that which is paid savers for their use'.

Appendix I (III) of his *Éléments d'économie pure*, where he deals with 'l'échange d'épargnes contre capitaux neufs', argues expressly that, corresponding to each possible rate of interest, there is a sum which individuals will save and also a sum which they will invest in new capital assets, that these two aggregates tend to equality with one another, and that the rate of interest is the variable which brings them to equality; so that the rate of interest is fixed at the point where saving, which represents the supply of new capital, is equal to the demand for it. Thus he is strictly in the classical tradition.

Certainly the ordinary man—banker, civil servant or politician—brought up on the traditional theory, and the trained economist also, has carried away with him the idea that whenever an individual performs an act of saving he has done something which automatically brings down the rate of interest, that this automatically stimulates the output of capital, and that the fall in the rate of interest is just so much as is necessary to stimulate the output of capital to an extent which is equal to the increment of saving; and, further, that this is a self-regulatory process of adjustment which takes place without the necessity for any special intervention or grandmotherly care on the part of the monetary authority. Similarly—and this is an even more general belief, even to-day—each additional act of investment will necessarily raise the rate of interest, if it is not offset by a change in the readiness to save.

Now the analysis of the previous chapters will have made it plain that this account of the matter must be erroneous. In tracing to its source the reason for the difference of opinion, let us, however, begin with the matters which are agreed.

Unlike the neo-classical school, who believe that saving and investment can be actually unequal, the classical school proper has accepted the view that they are equal. Marshall, for example, surely believed, although he did not expressly say so, that aggregate saving and aggregate investment are necessarily equal. Indeed, most members of the classical school carried this belief much too far; since they held that every act of increased saving by an individual necessarily brings into existence a corresponding act of increased investment. Nor is there any material difference, relevant in this context, between my schedule of the marginal efficiency of capital or investment demand-schedule and the demand curve for capital contemplated by some of the classical writers who have been quoted above. When we come to the propensity to consume and its corollary the propensity to save, we are nearer to a difference of opinion, owing to the emphasis which they have placed on the influence of the rate of interest on the propensity to save. But they would, presumably, not wish to deny that the level of income also has an important influence on the amount saved; whilst I, for my part, would not deny that the rate of interest may perhaps have an influence (though perhaps not of the kind which they suppose) on the amount saved *out of a given income*. All these points of agreement can be summed up in a proposition which the classical school would accept and I should not dispute; namely, that, if the level of income is assumed to be given, we can infer that the current rate of interest must lie at the point where the demand curve for capital corresponding to different rates of interest cuts the curve of the amounts saved out of the given income corresponding to different rates of interest.

But this is the point at which definite error creeps into the classical theory. If the classical school merely inferred from the above proposition that, given the demand curve for capital and the influence of changes in the rate of interest on the readiness to save out of given incomes, the level of income and the rate of interest must be uniquely correlated, there would be nothing to quarrel with. Moreover, this proposition would lead naturally to another proposition which embodies an important truth; namely, that, if the rate of interest is given as well as the demand curve for capital and the influence of the rate of interest on the readiness to save out of given levels of income, the level of income must be the factor which brings the amount saved to equality with the amount invested. But, in fact, the classical theory not merely neglects the influence of changes in the level of income, but involves formal error.

For the classical theory, as can be seen from the above quotations, assumes that it can then proceed to consider the effect on the rate of interest of (e.g.) a shift in the demand curve for capital, without abating or modifying its assumption as to the amount of the given income out of which the savings are to be made. The independent variables of the classical theory of the rate of interest are the demand curve for capital and the influence of the rate of interest on the amount saved out of a given income; and when (e.g.) the demand curve for capital shifts, the new rate of interest, according to this theory, is given by the point of intersection between the new demand curve for capital and the curve relating the rate of interest to the amounts which will be saved out of the given income. The classical theory of the rate of interest seems to suppose that, if the demand curve for capital shifts or if the curve relating the rate of interest to the amounts saved out of a given income shifts or if both these curves shift, the new rate of interest will be given by the point of intersection of the new positions of the two curves. But this is a nonsense theory. For the assumption that income is constant is inconsistent with the assumption that these two curves can shift independently of one another. If either of them shift, then, in general, income will change; with the result that the whole schematism based on the assumption of a given income breaks down. The position could only be saved by some complicated assumption providing for an automatic change in the wage-unit of an amount just sufficient in its effect on liquidity-preference to establish a rate of interest which would just offset the supposed shift, so as to leave output at the same level as before. In fact, there is no hint to be found in the above writers as to the necessity for any such assumption; at the best it would be plausible only in relation to long-period equilibrium and could not form the basis of a short-period theory; and there is no ground for supposing it to hold even in the long-period. In truth, the classical theory has not been alive to the relevance of changes in the level of income or to the possibility of the level of income being actually a function of the rate of the investment.

The above can be illustrated by a diagram[1] as follows:

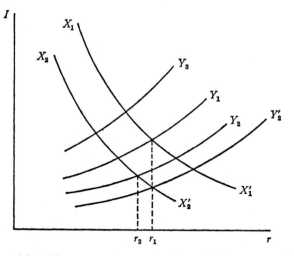

In this diagram the amount of investment (or saving) I is measured vertically, and the rate of interest r horizontally. $X_1 X_1'$ is the first position of the investment demand-schedule, and $X_2 X_2'$ is a second position of this curve. The curve Y_1 relates the

[1] This diagram was suggested to me by Mr R. F. Harrod. Cf. also a partly similar schematism by Mr D. H. Robertson, *Economic Journal*, December 1934, p. 652.

amounts saved out of an income Y_1 to various levels of the rate of interest, the curves Y_2, Y_3, etc., being the corresponding curves for levels of income Y_2, Y_3, etc. Let us suppose that the curve Y_1 is the Y-curve consistent with an investment demand-schedule $X_1 X_1'$ and a rate of interest r_1. Now if the investment demand-schedule shifts from $X_1 X_1'$ to $X_2 X_2'$, income will, in general, shift also. But the above diagram does not contain enough *data* to tell us what its new value will be; and, therefore, not knowing which is the appropriate Y-curve, we do not know at what point the new investment demand-schedule will cut it. If, however, we introduce the state of liquidity-preference and the quantity of money and these between them tell us that the rate of interest is r_2, then the whole position becomes determinate. For the Y-curve which intersects $X_2 X_2'$ at the point vertically above r_2, namely, the curve Y_2, will be the appropriate curve. Thus the X-curve and the Y-curves tell us nothing about the rate of interest. They only tell us what income will be, if from some other source we can say what the rate of interest is. If nothing has happened to the state of liquidity-preference and the quantity of money, so that the rate of interest is unchanged, then the curve Y_2' which intersects the new investment demand-schedule vertically below the point where the curve Y_1 intersected the old investment demand-schedule will be the appropriate Y-curve, and Y_2' will be the new level of income.

Thus the functions used by the classical theory, namely, the response of investment and the response of the amount saved out of a given income to change in the rate of interest, do not furnish material for a theory of the rate of interest; but they could be used to tell us what the level of income will be, given (from some other source) the rate of interest; and, alternatively, what the rate of interest will have to be, if the level of income is to be maintained at a given figure (e.g. the level corresponding to full employment).

The mistake originates from regarding interest as the reward for waiting as such, instead of as the reward for not-hoarding; just as the rates of return on loans or investments involving different degrees of risk, are quite properly regarded as the reward, not of waiting as such, but of running the risk. There is, in truth, no sharp line between these and the so-called 'pure' rate of interest, all of them being the reward for running the risk of uncertainty of one kind or another. Only in the event of money being used solely for transactions and never as a store of value, would a different theory become appropriate.[1]

There are, however, two familiar points which might, perhaps, have warned the classical school that something was wrong. In the first place, it has been agreed, at any rate since the publication of Professor Cassel's *Nature and Necessity of Interest*, that it is not certain that the sum saved out of a given income necessarily increases when the rate of interest is increased; whereas no one doubts that the investment demand-schedule falls with a rising rate of interest. But if the Y-curves and the X-curves both fall as the rate of interest rises, there is no guarantee that a given Y-curve will intersect a given X-curve anywhere at all. This suggests that it cannot be the Y-curve and the X-curve alone which determine the rate of interest.

In the second place, it has been usual to suppose that an increase in the quantity of money has a tendency to reduce the rate of interest, at any rate in the first instance and in the short period. Yet no reason has been given why a change in the quantity of money should affect either the investment demand-schedule or the readiness to save out of a given income. Thus the classical school have had quite a different theory of the rate of interest in volume I dealing with the theory of value from what they have had in volume II dealing with the theory of money. They have seemed un-

[1] Cf. chapter 17 below.

disturbed by the conflict and have made no attempt, so far as I know, to build a bridge between the two theories. The classical school proper, that is to say; since it is the attempt to build a bridge on the part of the neo-classical school which has led to the worst muddles of all. For the latter have inferred that there must be *two* sources of supply to meet the investment demand-schedule; namely, savings proper, which are the savings dealt with by the classical school, *plus* the sum made available by any increase in the quantity of money (this being balanced by some species of levy on the public, called 'forced saving' or the like). This leads on to the idea that there is a 'natural' or 'neutral'[1] or 'equilibrium' rate of interest, namely, that rate of interest which equates investment to classical savings proper without any addition from 'forced savings'; and finally to what, assuming they are on the right track at the start, is the most obvious solution of all, namely, that, if the quantity of money could only be kept *constant* in all circumstances, none of these complications would arise, since the evils supposed to result from the supposed excess of investment over savings proper would cease to be possible. But at this point we are in deep water. 'The wild duck has dived down to the bottom—as deep as she can get—and bitten fast hold of the weed and tangle and all the rubbish that is down there, and it would need an extraordinarily clever dog to dive after and fish her up again.'

Thus the traditional analysis is faulty because it has failed to isolate correctly the independent variables of the system. Saving and investment are the determinates of the system, not the determinants. They are the twin results of the system's determinants, namely, the propensity to consume, the schedule of the marginal efficiency of capital and the rate of interest. These

---

[1] The 'neutral' rate of interest of contemporary economists is different both from the 'natural' rate of Böhm–Bawerk and from the 'natural' rate of Wicksell.

determinants are, indeed, themselves complex and each is capable of being affected by prospective changes in the others. But they remain independent in the sense that their values cannot be inferred from one another. The traditional analysis has been aware that saving depends on income but it has overlooked the fact that income depends on investment, in such fashion that, when investment changes, income must necessarily change in just that degree which is necessary to make the change in saving equal to the change in investment.

Nor are those theories more successful which attempt to make the rate of interest depend on 'the marginal efficiency of capital'. It is true that in equilibrium the rate of interest will be equal to the marginal efficiency of capital, since it will be profitable to increase (or decrease) the current scale of investment until the point of equality has been reached. But to make this into a theory of the rate of interest or to derive the rate of interest from it involves a circular argument, as Marshall discovered after he had got half-way into giving an account of the rate of interest along these lines.[1] For the 'marginal efficiency of capital' partly depends on the scale of current investment, and we must already know the rate of interest before we can calculate what this scale will be. The significant conclusion is that the output of new investment will be pushed to the point at which the marginal efficiency of capital becomes equal to the rate of interest; and what the schedule of the marginal efficiency of capital tells us, is, not what the rate of interest is, but the point to which the output of new investment will be pushed, given the rate of interest.

The reader will readily appreciate that the problem here under discussion is a matter of the most fundamental theoretical significance and of overwhelming practical importance. For the economic principle,

[1] See the Appendix to this chapter.

on which the practical advice of economists has been almost invariably based, has assumed, in effect, that, *cet. par.*, a decrease in spending will tend to lower the rate of interest and an increase in investment to raise it. But if what these two quantities determine is, not the rate of interest, but the aggregate volume of employment, then our outlook on the mechanism of the economic system will be profoundly changed. A decreased readiness to spend will be looked on in quite a different light if, instead of being regarded as a factor which will, *cet. par.*, increase investment, it is seen as a factor which will, *cet. par.*, diminish employment.

Appendix to Chapter 14

APPENDIX ON THE RATE OF INTEREST IN MARSHALL'S *PRIN-CIPLES OF ECONOMICS*, RICARDO'S *PRINCIPLES OF POLITICAL ECONOMY*, AND ELSEWHERE

I

There is no consecutive discussion of the rate of interest in the works of Marshall, Edgeworth or Professor Pigou,—nothing more than a few *obiter dicta*. Apart from the passage already quoted above (p. 139) the only important clues to Marshall's position on the rate of interest are to be found in his *Principles of Economics* (6th edn.), Book VI. p. 534 and p. 593, the gist of which is given by the following quotations:

'Interest, being the price paid for the use of capital in any market, tends towards an equilibrium level such that the aggregate demand for capital in that market, at that rate of interest, is equal to the aggregate stock[1] forthcoming there at that rate. If the market, which we are considering, is a small one—say a single town, or a single trade in a progressive country—an increased demand for capital in it will be promptly met by an increased supply drawn from surrounding districts or trades. But if we are considering the whole world, or even the whole of a large country, as one market for capital, we cannot regard the aggregate supply of it as altered quickly and to a considerable extent by a change in the rate of interest. For the general fund of capital is the product of labour and waiting; and the extra

[1] It is to be noticed that Marshall uses the word 'capital' not 'money' and the word 'stock' not 'loans'; yet interest is a payment for borrowing *money*, and 'demand for capital' in this context should mean 'demand for loans of money for the purpose of buying a stock of capital-goods'. But the equality between the stock of capital-goods offered and the stock demanded will be brought about by the *prices* of capital-goods, not by the rate of interest. It is equality between the demand and supply of loans of money, i.e. of debts, which is brought about by the rate of interest.

work,[1] and the extra waiting, to which a rise in the rate of interest would act as an incentive, would not quickly amount to much, as compared with the work and waiting, of which the total existing stock of capital is the result. An extensive increase in the demand for capital in general will therefore be met for a time not so much by an increase of supply, as by a rise in the rate of interest;[2] which will cause capital to withdraw itself partially from those uses in which its marginal utility is lowest. It is only slowly and gradually that the rise in the rate of interest will increase the total stock of capital' (p. 534).

'It cannot be repeated too often that the phrase "the rate of interest" is applicable to old investments of capital only in a very limited sense.'[3] For instance, we may perhaps estimate that a trade capital of some seven thousand millions is invested in the different trades of this country at about 3 per cent net interest. But such a method of speaking, though convenient and justifiable for many purposes, is not accurate. What ought

---

[1] This assumes that income is *not* constant. But it is not obvious in what way a rise in the rate of interest will lead to 'extra work'. Is the suggestion that a rise in the rate of interest is to be regarded, by reason of its increasing the attractiveness of working in order to save, as constituting a sort of increase in real wages which will induce the factors of production to work for a lower wage? This is, I think in Mr D. H. Robertson's mind in a similar context. Certainly this 'would not quickly amount to much'; and an attempt to explain the actual fluctuations in the amount of investment by means of this factor would be most unplausible, indeed absurd. My rewriting of the latter half of this sentence would be: 'and if an extensive increase in the demand for capital in general, due to an increase in the schedule of the marginal efficiency of capital, is *not* offset by a rise in the rate of interest, the extra employment and the higher level of income, which will ensue as a result of the increased production of capital-goods, will lead to an amount of extra waiting which in terms of money will be exactly equal to the value of the current increment of capital-goods and will, therefore, precisely provide for it'.

[2] Why not by a rise in the supply price of capital-goods? Suppose, for example, that the 'extensive increase in the demand for capital in general' is due to a *fall* in the rate of interest. I would suggest that the sentence should be rewritten: 'In so far, therefore, as the extensive increase in the demand for capital-goods cannot be immediately met by an increase in the total stock, it will have to be held in check for the time being by a rise in the supply price of capital-goods sufficient to keep the marginal efficiency of capital in equilibrium with the rate of interest without there being any material change in the scale of investment; meanwhile (as always) the factors of production adapted for the output of capital-goods will be used in producing those capital-goods of which the marginal efficiency is greatest in the new conditions.'

[3] In fact we cannot speak of it at all. We can only properly speak of the rate of interest on *money* borrowed for the purpose of purchasing investments of capital, new or old (or for any other purpose).

to be said is that, taking the rate of net interest on the investments of new capital in each of those trades [i.e. on marginal investments] to be about 3 per cent; then the aggregate net income rendered by the whole of the trade-capital invested in the various trades is such that, if capitalised at 33 years' purchase (that is, on the basis of interest at 3 per cent), it would amount to some seven thousand million pounds. For the value of the capital already invested in improving land or erecting a building, in making a railway or a machine, is the aggregate discounted value of its estimated future net incomes [or quasi-rents]; and if its prospective income-yielding power should diminish, its value would fall accordingly and would be the capitalised value of that smaller income after allowing for depreciation' (p. 593).

In his *Economics of Welfare* (3rd edn.), p. 163, Professor Pigou writes: 'The nature of the service of "waiting" has been much misunderstood. Sometimes it has been supposed to consist in the provision of money, sometimes in the provision of time, and, on both suppositions, it has been argued that no contribution whatever is made by it to the dividend. Neither supposition is correct. "Waiting" simply means postponing consumption which a person has power to enjoy immediately, thus allowing resources, which might have been destroyed, to assume the form of production instruments[1]...The unit of "waiting" is, therefore, the use of a given quantity of resources[2]—for example, labour or machinery—for a given time...In more general terms we may say that the unit of waiting is a year-value-unit, or, in the simpler, if less accurate, language of Dr Cassel, a year-pound...A caution may be added against the common view that the amount of capital accumulated in any year is necessarily equal to the amount of "savings" made in it. This is not so, even when savings are interpreted to mean net savings, thus eliminating the savings of one man that are lent to increase the consumption of another, and when temporary accumulations of *unused* claims upon services in the form of bank-money are ignored; for many savings which are meant to become capital

---

[1] Here the wording is ambiguous as to whether we are to infer that the postponement of consumption *necessarily* has this effect, or whether it merely releases resources which are then either unemployed or used for investment according to circumstances.

[2] Not, be it noted, the amount of money which the recipient of income might, but does not, spend on consumption; so that the reward of waiting is not interest but quasi-rent. This sentence seems to imply that the released resources are necessarily *used*. For what is the reward of waiting if the released resources are left unemployed?

in fact fail of their purpose through misdirection into wasteful uses.'[1]

Professor Pigou's only significant reference to what determines the rate of interest is, I think, to be found in his *Industrial Fluctuations* (1st edn.), pp. 251-3, where he controverts the view that the rate of interest, being determined by the general conditions of demand and supply of real capital, lies outside the central or any other bank's control. Against this view he argues that: 'When bankers create more credit for business men, they make, in their interest, subject to the explanations given in chapter xiii. of part i.,[2] a forced levy of real things from the public, thus increasing the stream of real capital available for them, and causing a fall in the real rate of interest on long and short loans alike. It is true, in short, that the bankers' rate for money is bound by a mechanical tie to the real rate of interest on long loans; but it is not true that this real rate is determined by conditions wholly outside bankers' control.'

My running comments on the above have been made in the footnotes. The perplexity which I find in Marshall's account of the matter is fundamentally due, I think, to the incursion of the concept 'interest', which belongs to a monetary economy, into a treatise which takes no account of money. 'Interest' has really no business to turn up at all in Marshall's *Principles of Economics*,—it belongs to another branch of the subject.

[1] We are not told in this passage whether net savings would or would not be equal to the increment of capital, if we were to ignore misdirected investment but were to take account of 'temporary accumulations of *unused* claims upon services in the form of bank-money'. But in *Industrial Fluctuations* (p. 22) Prof. Pigou makes it clear that such accumulations have no effect on what he calls 'real savings'.

[2] This reference (*op. cit.* pp. 129-134) contains Prof. Pigou's view as to the amount by which a new credit creation by the banks increases the stream of real capital available for entrepreneurs. In effect he attempts to deduct 'from the floating credit handed over to business men through credit creations the floating capital which would have been contributed in other ways if the banks had not been there'. After these deductions have been made, the argument is one of deep obscurity. To begin with, the rentiers have an income of 1500, of which they consume 500 and save 1000; the act of credit creation reduces their income to 1300, of which they consume $500-x$ and save $800+x$; and x, Prof. Pigou concludes, represents the net increase of capital made available by the act of credit creation. Is the entrepreneurs' *income* supposed to be swollen by the amount which they *borrow* from the banks (after making the above deductions)? Or is it swollen by the amount, i.e. 200, by which the rentiers' income is reduced? In either case, are they supposed to save the whole of it? Is the increased investment equal to the credit creations *minus* the deductions? Or is it equal to x? The argument seems to stop just where it should begin.

Professor Pigou, conformably with his other tacit assumptions, leads us (in his *Economics of Welfare*) to infer that the unit of waiting is the same as the unit of current investment and that the reward of waiting is quasi-rent, and practically never mentions interest,—which is as it should be. Nevertheless these writers are not dealing with a non-monetary economy (if there is such a thing). They quite clearly presume that money is used and that there is a banking system. Moreover, the rate of interest scarcely plays a larger part in Professor Pigou's *Industrial Fluctuations* (which is mainly a study of fluctuations in the marginal efficiency of capital) or in his *Theory of Unemployment* (which is mainly a study of what determines changes in the volume of employment, assuming that there is no involuntary unemployment) than in his *Economics of Welfare*.

II

The following from his *Principles of Political Economy* (p. 511) puts the substance of Ricardo's theory of the rate of interest:

'The interest of money is not regulated by the rate at which the Bank will lend, whether it be 5, 3 or 2 per cent., but by the rate of profit which can be made by the employment of capital, and which is totally independent of the quantity or of the value of money. Whether the Bank lent one million, ten millions, or a hundred millions, they would not permanently alter the market rate of interest; they would alter only the value of the money which they thus issued. In one case, ten or twenty times more money might be required to carry on the same business than what might be required in the other. The applications to the Bank for money, then, depend on the comparison between the rate of profits that may be made by the employment of it, and the rate at which they are willing to lend it. If they charge less than the market rate of interest, there is no amount of money which they might not lend;—if they charge more than that rate, none but spendthrifts and prodigals would be found to borrow of them.'

This is so clear-cut that it affords a better starting-point for a discussion than the phrases of later writers who, without really departing from the essence of the Ricardian doctrine, are nevertheless sufficiently uncomfortable about it to seek refuge in haziness. The above is, of course, as always with Ricardo, to be interpreted as a long-period doctrine, with the emphasis on the word 'permanently' half-way through the passage; and it is interesting to consider the assumptions required to validate it.

Once again the assumption required is the usual classical assumption, that there is always full employment; so that, assuming no change in the supply curve of labour in terms of product, there is only one possible level of employment in long-period equilibrium. On this assumption with the usual *ceteris paribus*, i.e. no change in psychological propensities and expectations other than those arising out of a change in the quantity of money, the Ricardian theory is valid, in the sense that on these suppositions there is only one rate of interest which will be compatible with full employment in the long period. Ricardo and his successors overlook the fact that even in the long period the volume of employment is not necessarily full but is capable of varying, and that to every banking policy there corresponds a different long-period level of employment; so that there are a number of positions of long-period equilibrium corresponding to different conceivable interest policies on the part of the monetary authority.

If Ricardo had been content to present his argument solely as applying to any given quantity of money created by the monetary authority, it would still have been correct on the assumption of flexible money-wages. If, that is to say, Ricardo had argued that it would make no permanent alteration to the rate of interest whether the quantity of money was fixed by the monetary authority at ten millions or at a hundred millions, his conclusion would hold. But if by the policy of the monetary authority we mean the terms on which it will increase or decrease the quantity of money, i.e. the rate of interest at which it will, either by a change in the volume of discounts or by open-market operations, increase or decrease its assets—which is what Ricardo expressly does mean in the above quotation—then it is not the case either that the policy of the monetary authority is nugatory or that only one policy is compatible with long-period equilibrium; though in the extreme case where money-wages are assumed to fall without limit in face of involuntary unemployment through a futile competition for employment between the unemployed labourers, there will, it is true, be only two possible long-period positions—full employment and the level of employment corresponding to the rate of interest at which liquidity-preference becomes absolute (in the event of this being less than full employment). Assuming flexible money-wages, the quantity of money as such is, indeed, nugatory in the long period; but the terms on which the monetary authority will change the quantity of money enters as a real determinant into the economic scheme.

It is worth adding that the concluding sentences of the quotation suggest that Ricardo was overlooking the possible changes in the marginal efficiency of capital according to the amount invested. But this again can be interpreted as another example of his greater internal consistency compared with his successors. For if the quantity of employment and the psychological propensities of the community are taken as given, there is in fact only one possible rate of accumulation of capital and, consequently, only one possible value for the marginal efficiency of capital. Ricardo offers us the supreme intellectual achievement, unattainable by weaker spirits, of adopting a hypothetical world remote from experience as though it were the world of experience and then living in it consistently. With most of his successors common sense cannot help breaking in—with injury to their logical consistency.

III

A peculiar theory of the rate of interest has been propounded by Professor von Mises and adopted from him by Professor Hayek and also, I think, by Professor Robbins; namely, that changes in the rate of interest can be identified with changes in the relative price levels of consumption-goods and capital-goods.[1] It is not clear how this conclusion is reached. But the argument seems to run as follows. By a somewhat drastic simplification the marginal efficiency of capital is taken as measured by the ratio of the supply price of new consumers' goods to the supply price of new producers' goods.[2] This is then identified with the rate of interest. The fact is called to notice that a fall in the rate of interest is favourable to investment. *Ergo*, a fall in the ratio of the price of consumers' goods to the price of producer's goods is favourable to investment.

By this means a link is established between increased saving by an individual and increased aggregate investment. For it is common gound that increased individual saving will cause a fall in the price of consumers' goods, and, quite possibly, a

[1] *The Theory of Money and Credit*, p. 339 *et passim*, particularly p. 363.
[2] If we are in long-period equilibrium, special assumptions might be devised on which this could be justified. But when the prices in question are the prices prevailing in slump conditions, the simplification of supposing that the entrepreneur will, in forming his expectations, assume these prices to be permanent, is certain to be misleading. Moreover, if he does, the prices of the existing stock of producers' goods will fall in the same proportion as the prices of consumers' goods.

greater fall than in the price of producers' goods; hence, according to the above reasoning, it means a reduction in the rate of interest which will stimulate investment. But, of course, a lowering of the marginal efficiency of particular capital assets, and hence a lowering of the schedule of the marginal efficiency of capital in general, has exactly the opposite effect to what the above argument assumes. For investment is stimulated either by a *raising* of the schedule of the marginal efficiency or by a *lowering* of the rate of interest. As a result of confusing the marginal efficiency of capital with the rate of interest, Professor von Mises and his disciples have got their conclusions exactly the wrong way round. A good example of a confusion along these lines is given by the following passage by Professor Alvin Hansen:[1] 'It has been suggested by some economists that the net effect of reduced spending will be a lower price level of consumers' goods than would otherwise have been the case, and that, in consequence, the stimulus to investment in fixed capital would thereby tend to be minimised. This view is, however, incorrect and is based on a confusion of the effect on capital formation of (1) higher or lower prices of consumers' goods, and (2) a change in the rate of interest. It is true that in consequence of the decreased spending and increased saving, consumers' prices would be low relative to the prices of producers' goods. But this, in effect, means a lower rate of interest, and a lower rate of interest stimulates an expansion of capital investment in fields which at higher rates would be unprofitable.'

[1] *Economic Reconstruction*, p. 233.

Chapter 15

THE PSYCHOLOGICAL AND
BUSINESS INCENTIVES TO
LIQUIDITY

I

We must now develop in more detail the analysis of the motives to liquidity-preference which were introduced in a preliminary way in chapter 13. The subject is substantially the same as that which has been sometimes discussed under the heading of the demand for money. It is also closely connected with what is called the income-velocity of money;—for the income-velocity of money merely measures what proportion of their incomes the public chooses to hold in cash, so that an increased income-velocity of money may be a symptom of a decreased liquidity-preference. It is not the same thing, however, since it is in respect of his stock of accumulated savings, rather than of his income, that the individual can exercise his choice between liquidity and illiquidity. And, anyhow, the term 'income-velocity of money' carries with it the misleading suggestion of a presumption in favour of the demand for money as a whole being proportional, or having some determinate relation, to income, whereas this presumption should apply, as we shall see, only to a *portion* of the public's cash holdings; with the result that it overlooks the part played by the rate of interest.

In my *Treatise on Money* I studied the total demand for money under the headings of income-deposits, business-deposits, and savings-deposits, and I need not repeat here the analysis which I gave in chapter 3 of that book. Money held for each of the three purposes forms, nevertheless, a single pool, which the holder is under no necessity to segregate into three water-tight compartments; for they need not be sharply divided even in his own mind, and the same sum can be held primarily for one purpose and secondarily for another. Thus we can—equally well, and, perhaps, better— consider the individual's aggregate demand for money in given circumstances as a single decision, though the composite result of a number of different motives.

In analysing the motives, however, it is still convenient to classify them under certain headings, the first of which broadly corresponds to the former classification of income-deposits and business-deposits, and the two latter to that of savings-deposits. These I have briefly introduced in chapter 13 under the headings of the transactions-motive, which can be further classified as the income-motive and the business-motive, the precautionary-motive and the speculative-motive.

(i) *The Income-motive.* One reason for holding cash is to bridge the interval between the receipt of income and its disbursement. The strength of this motive in inducing a decision to hold a given aggregate of cash will chiefly depend on the amount of income and the normal length of the interval between its receipt and its disbursement. It is in this connection that the concept of the income-velocity of money is strictly appropriate.

(ii) *The Business-motive.* Similarly, cash is held to bridge the interval between the time of incurring business costs and that of the receipt of the sale-proceeds; cash held by dealers to bridge the interval between purchase and realisation being included under this heading. The strength of this demand will chiefly depend on the value of current output (and hence on current income), and on the number of hands through which output passes.

(iii) *The Precautionary-motive.* To provide for contingencies requiring sudden expenditure and for unforeseen opportunities of advantageous purchases, and also to hold an asset of which the value is fixed in terms of money to meet a subsequent liability fixed in terms of money, are further motives for holding cash.

The strength of all these three types of motive will partly depend on the cheapness and the reliability of methods of obtaining cash, when it is required, by some form of temporary borrowing, in particular by overdraft or its equivalent. For there is no necessity to hold idle cash to bridge over intervals if it can be obtained without difficulty at the moment when it is actually required. Their strength will also depend on what we may term the relative cost of holding cash. If the cash can only be retained by forgoing the purchase of a profitable asset, this increases the cost and thus weakens the motive towards holding a given amount of cash. If deposit interest is earned or if bank charges are avoided by holding cash, this decreases the cost and strengthens the motive. It may be, however, that this is likely to be a minor factor except where large changes in the cost of holding cash are in question.

(iv) There remains the *Speculative-motive.* This needs a more detailed examination than the others, both because it is less well understood and because it is particularly important in transmitting the effects of a *change* in the quantity of money.

In normal circumstances the amount of money required to satisfy the transactions-motive and the precautionary-motive is mainly a resultant of the general activity of the economic system and of the level of money-income. But it is by playing on the speculative-motive that monetary management (or, in the absence of management, chance changes in the quantity of money) is brought to bear on the economic system. For the demand for money to satisfy the former motives is generally irresponsive to any influence except the actual occurrence of a change in the general economic activity and the level of incomes; whereas experience indicates that the aggregate demand for money to satisfy the speculative-motive usually shows a continuous response to gradual changes in the rate of interest, i.e. there is a continuous curve relating changes in the demand for money to satisfy the speculative motive and changes in the rate of interest as given by changes in the prices of bonds and debts of various maturities.

Indeed, if this were not so, 'open market operations' would be impracticable. I have said that experience indicates the continuous relationship stated above, because in normal circumstances the banking system is in fact always able to purchase (or sell) bonds in exchange for cash by bidding the price of bonds up (or down) in the market by a modest amount; and the larger the quantity of cash which they seek to create (or cancel) by purchasing (or selling) bonds and debts, the greater must be the fall (or rise) in the rate of interest. Where, however, (as in the United States, 1933–1934) open-market operations have been limited to the purchase of very short-dated securities, the effect may, of course, be mainly confined to the very short-term rate of interest and have but little reaction on the much more important long-term rates of interest.

In dealing with the speculative-motive it is, however, important to distinguish between the changes in the rate of interest which are due to changes in the supply of money available to satisfy the speculative-motive, without there having been any change in the liquidity function, and those which are primarily due to changes in expectation affecting the liquidity function itself. Open-market operations may, indeed, influence the rate of interest through both channels; since they may not only change the volume of

197

money, but may also give rise to changed expectations concerning the future policy of the central bank or of the government. Changes in the liquidity function itself, due to a change in the news which causes revision of expectations, will often be discontinuous, and will, therefore, give rise to a corresponding discontinuity of change in the rate of interest. Only, indeed, in so far as the change in the news is differently interpreted by different individuals or affects individual interests differently will there be room for any increased activity of dealing in the bond market. If the change in the news affects the judgment and the requirements of everyone in precisely the same way, the rate of interest (as indicated by the prices of bonds and debts) will be adjusted forthwith to the new situation without any market transactions being necessary.

Thus, in the simplest case, where everyone is similar and similarly placed, a change in circumstances or expectations will not be capable of causing any displacement of money whatever;—it will simply change the rate of interest in whatever degree is necessary to offset the desire of each individual, felt at the previous rate, to change his holding of cash in response to the new circumstances or expectations; and, since everyone will change his ideas as to the rate which would induce him to alter his holdings of cash in the same degree, no transactions will result. To each set of circumstances and expectations there will correspond an appropriate rate of interest, and there will never be any question of anyone changing his usual holdings of cash.

In general, however, a change in circumstances or expectations will cause some realignment in individual holdings of money;—since, in fact, a change will influence the ideas of different individuals differently by reasons partly of differences in environment and the reason for which money is held and partly of differences in knowledge and interpretation of the new situation. Thus the new equilibrium rate of interest will be associated with a redistribution of money-holdings. Nevertheless it is the change in the rate of interest, rather than the redistribution of cash, which deserves our main attention. The latter is incidental to individual differences, whereas the essential phenomenon is that which occurs in the simplest case. Moreover, even in the general case, the shift in the rate of interest is usually the most prominent part of the reaction to a change in the news. The movement in bond-prices is, as the newspapers are accustomed to say, 'out of all proportion to the activity of dealing';—which is as it should be, in view of individuals being much more similar than they are dissimilar in their reaction to news.

II

Whilst the amount of cash which an individual decides to hold to satisfy the transactions-motive and the precautionary-motive is not entirely independent of what he is holding to satisfy the speculative-motive, it is a safe first approximation to regard the amounts of these two sets of cash-holdings as being largely independent of one another. Let us, therefore, for the purposes of our further analysis, break up our problem in this way.

Let the amount of cash held to satisfy the transactions- and precautionary-motives be M_1, and the amount held to satisfy the speculative-motive be M_2. Corresponding to these two compartments of cash, we then have two liquidity functions L_1 and L_2. L_1 mainly depends on the level of income, whilst L_2 mainly depends on the relation between the current rate of interest and the state of expectation. Thus

$$M = M_1 + M_2 = L_1(Y) + L_2(r),$$

where L_1 is the liquidity function corresponding to

an income Y, which determines M_1, and L_2 is the liquidity function of the rate of interest r, which determines M_2. It follows that there are three matters to investigate: (i) the relation of changes in M to Y and r, (ii) what determines the shape of L_1, (iii) what determines the shape of L_2.

(i) The relation of changes in M to Y and r depends, in the first instance, on the way in which changes in M come about. Suppose that M consists of gold coins and that changes in M can only result from increased returns to the activities of gold-miners who belong to the economic system under examination. In this case changes in M are, in the first instance, directly associated with changes in Y, since the new gold accrues as someone's income. Exactly the same conditions hold if changes in M are due to the government printing money wherewith to meet its current expenditure;—in this case also the new money accrues as someone's income. The new level of income, however, will not continue sufficiently high for the requirements of M_1 to absorb the whole of the increase in M; and some portion of the money will seek an outlet in buying securities or other assets until r has fallen so as to bring about an increase in the magnitude of M_2 and at the same time to stimulate a rise in Y to such an extent that the new money is absorbed either in M_2 or in the M_1 which corresponds to the rise in Y caused by the fall in r. Thus at one remove this case comes to the same thing as the alternative case, where the new money can only be issued in the first instance by a relaxation of the conditions of credit by the banking system, so as to induce someone to sell the banks a debt or a bond in exchange for the new cash.

It will, therefore, be safe for us to take the latter case as typical. A change in M can be assumed to operate by changing r, and a change in r will lead to a new equilibrium partly by changing M_2 and partly by changing Y and therefore M_1. The division of the increment of cash between M_1 and M_2 in the new position of equilibrium will depend on the responses of investment to a reduction in the rate of interest and of income to an increase in investment.[1] Since Y partly depends on r, it follows that a given change in M has to cause a sufficient change in r for the resultant changes in M_1 and M_2 respectively to add up to the given change in M.

(ii) It is not always made clear whether the income-velocity of money is defined as the ratio of Y to M or as the ratio of Y to M_1. I propose, however, to take it in the latter sense. Thus if V is the income-velocity of money,

$$L_1(Y) = \frac{Y}{V} = M_1.$$

There is, of course, no reason for supposing that V is constant. Its value will depend on the character of banking and industrial organisation, on social habits, on the distribution of income between different classes and on the effective cost of holding idle cash. Nevertheless, if we have a short period of time in view and can safely assume no material change in any of these factors, we can treat V as nearly enough constant.

(iii) Finally there is the question of the relation between M_2 and r. We have seen in chapter 13 that *uncertainty* as to the future course of the rate of interest is the sole intelligible explanation of the type of liquidity-preference L_2 which leads to the holding of cash M_2. It follows that a given M_2 will not have a definite quantitative relation to a given rate of interest of r;—what matters is not the *absolute* level of r but the degree of its divergence from what is considered a fairly *safe* level of r, having regard to those calculations of probability which are being relied on. Nevertheless, there are two reasons for expecting that, in

---

[1] We must postpone to Book V the question of what will determine the character of the new equilibrium.

any given state of expectation, a fall in r will be associated with an increase in M_2. In the first place, if the general view as to what is a safe level of r is unchanged, every fall in r reduces the market rate relatively to the 'safe' rate and therefore increases the risk of illiquidity; and, in the second place, every fall in r reduces the current earnings from illiquidity, which are available as a sort of insurance premium to offset the risk of loss on capital account, by an amount equal to the difference between the *squares* of the old rate of interest and the new. For example, if the rate of interest on a long-term debt is 4 per cent, it is preferable to sacrifice liquidity unless on a balance of probabilities it is feared that the long-term rate of interest may rise faster than by 4 per cent of itself per annum, i.e. by an amount greater than o·16 per cent per annum. If, however, the rate of interest is already as low as 2 per cent, the running yield will only offset a rise in it of as little as o·04 per cent per annum. This, indeed, is perhaps the chief obstacle to a fall in the rate of interest to a very low level. Unless reasons are believed to exist why future experience will be very different from past experience, a long-term rate of interest of (say) 2 per cent leaves more to fear than to hope, and offers, at the same time, a running yield which is only sufficient to offset a very small measure of fear.

It is evident, then, that the rate of interest is a highly psychological phenomenon. We shall find, indeed, in Book V that it cannot be in equilibrium at a level *below* the rate which corresponds to full employment; because at such a level a state of true inflation will be produced, with the result that M_1 will absorb ever-increasing quantities of cash. But at a level *above* the rate which corresponds to full employment, the long-term market-rate of interest will depend, not only on the current policy of the monetary authority, but also on market expectations concerning its future policy. The short-term rate of interest is easily controlled by the monetary authority, both because it is not difficult to produce a conviction that its policy will not greatly change in the very near future, and also because the possible loss is small compared with the running yield (unless it is approaching vanishing point). But the long-term rate may be more recalcitrant when once it has fallen to a level which, on the basis of past experience and present expectations of *future* monetary policy, is considered 'unsafe' by representative opinion. For example, in a country linked to an international gold standard, a rate of interest lower than prevails elsewhere will be viewed with a justifiable lack of confidence; yet a domestic rate of interest dragged up to a parity with the *highest* rate (highest after allowing for risk) prevailing in any country belonging to the international system may be much higher than is consistent with domestic full employment.

Thus a monetary policy which strikes public opinion as being experimental in character or easily liable to change may fail in its objective of greatly reducing the long-term rate of interest, because M_2 may tend to increase almost without limit in response to a reduction of r below a certain figure. The same policy, on the other hand, may prove easily successful if it appeals to public opinion as being reasonable and practicable and in the public interest, rooted in strong conviction, and promoted by an authority unlikely to be superseded.

It might be more accurate, perhaps, to say that the rate of interest is a highly conventional, rather than a highly psychological, phenomenon. For its actual value is largely governed by the prevailing view as to what its value is expected to be. *Any* level of interest which is accepted with sufficient conviction as *likely* to be durable *will* be durable; subject, of course, in a changing society to fluctuations for all kinds of reasons round the expected normal. In particular, when M_1

is increasing faster than M, the rate of interest will rise, and *vice versa*. But it may fluctuate for decades about a level which is chronically too high for full employment;—particularly if it is the prevailing opinion that the rate of interest is self-adjusting, so that the level established by convention is thought to be rooted in objective grounds much stronger than convention, the failure of employment to attain an optimum level being in no way associated, in the minds either of the public or of authority, with the prevalence of an inappropriate range of rates of interest.

The difficulties in the way of maintaining effective demand at a level high enough to provide full employment, which ensue from the association of a conventional and fairly stable long-term rate of interest with a fickle and highly unstable marginal efficiency of capital, should be, by now, obvious to the reader.

Such comfort as we can fairly take from more encouraging reflections must be drawn from the hope that, precisely because the convention is not rooted in secure knowledge, it will not be always unduly resistant to a modest measure of persistence and consistency of purpose by the monetary authority. Public opinion can be fairly rapidly accustomed to a modest fall in the rate of interest and the conventional expectation of the future may be modified accordingly; thus preparing the way for a further movement—up to a point. The fall in the long-term rate of interest in Great Britain after her departure from the gold standard provides an interesting example of this;—the major movements were effected by a series of discontinuous jumps, as the liquidity function of the public, having become accustomed to each successive reduction, became ready to respond to some new incentive in the news or in the policy of the authorities.

III

We can sum up the above in the proposition that in any given state of expectation there is in the minds of the public a certain potentiality towards holding cash beyond what is required by the transactions-motive or the precautionary-motive, which will realise itself in actual cash-holdings in a degree which depends on the terms on which the monetary authority is willing to create cash. It is this potentiality which is summed up in the liquidity function L_2.

Corresponding to the quantity of money created by the monetary authority, there will, therefore, be *cet. par.* a determinate rate of interest or, more strictly, a determinate complex of rates of interest for debts of different maturities. The same thing, however, would be true of any other factor in the economic system taken separately. Thus this particular analysis will only be useful and significant in so far as there is some specially direct or purposive connection between changes in the quantity of money and changes in the rate of interest. Our reason for supposing that there is such a special connection arises from the fact that, broadly speaking, the banking system and the monetary authority are dealers in money and debts and not in assets or consumables.

If the monetary authority were prepared to deal both ways on specified terms in debts of all maturities, and even more so if it were prepared to deal in debts of varying degrees of risk, the relationship between the complex of rates of interest and the quantity of money would be direct. The complex of rates of interest would simply be an expression of the terms on which the banking system is prepared to acquire or part with debts; and the quantity of money would be the amount which can find a home in the possession of individuals who—after taking account of all relevant circumstances —prefer the control of liquid cash to parting with it in exchange for a debt on the terms indicated by the market rate of interest. Perhaps a complex offer by the central bank to buy and sell at stated prices gilt-edged bonds of all maturities, in place of the single bank rate for short-term bills, is the most important practical improvement which can be made in the technique of monetary management.

To-day, however, in actual practice, the extent to which the price of debts as fixed by the banking system is 'effective' in the market, in the sense that it governs the actual market-price, varies in different systems. Sometimes the price is more effective in one direction than in the other; that is to say, the banking system may undertake to purchase debts at a certain price but not necessarily to sell them at a figure near enough to its buying-price to represent no more than a dealer's turn, though there is no reason why the price should not be made effective both ways with the aid of open-market operations. There is also the more important qualification which arises out of the monetary authority not being, as a rule, an equally willing dealer in debts of all maturities. The monetary authority often tends in practice to concentrate upon short-term debts and to leave the price of long-term debts to be influenced by belated and imperfect reactions from the price of short-term debts;—though here again there is no reason why they need do so. Where these qualifications operate, the directness of the relation between the rate of interest and the quantity of money is correspondingly modified. In Great Britain the field of deliberate control appears to be widening. But in applying this theory in any particular case allowance must be made for the special characteristics of the method actually employed by the monetary authority. If the monetary authority deals only in short-term debts, we have to consider what influence the price, actual and prospective, of short-term debts exercises on debts of longer maturity.

Thus there are certain limitations on the ability of the monetary authority to establish any given complex of rates of interest for debts of different terms and risks, which can be summed up as follows:

(1) There are those limitations which arise out of the monetary authority's own practices in limiting its willingness to deal to debts of a particular type.

(2) There is the possibility, for the reasons discussed above, that, after the rate of interest has fallen to a certain level, liquidity-preference may become virtually absolute in the sense that almost everyone prefers cash to holding a debt which yields so low a rate of interest. In this event the monetary authority would have lost effective control over the rate of interest. But whilst this limiting case might become practically important in future, I know of no example of it hitherto. Indeed, owing to the unwillingness of most monetary authorities to deal boldly in debts of long term, there has not been much opportunity for a test. Moreover, if such a situation were to arise, it would mean that the public authority itself could borrow through the banking system on an unlimited scale at a nominal rate of interest.

(3) The most striking examples of a complete breakdown of stability in the rate of interest, due to the liquidity function flattening out in one direction or the other, have occurred in very abnormal circumstances. In Russia and Central Europe after the war a currency crisis or flight from the currency was experienced, when no one could be induced to retain holdings either of money or of debts on any terms whatever, and even a high and rising rate of interest was unable to keep pace with the marginal efficiency of capital (especially of stocks of liquid goods) under the influence of the expectation of an ever greater fall in the value of money; whilst in the United States at certain dates in 1932 there was a crisis of the opposite kind—a financial crisis or crisis of liquidation, when

scarcely anyone could be induced to part with holdings of money on any reasonable terms.

(4) There is, finally, the difficulty discussed in section IV of chapter 11, p. 144, in the way of bringing the effective rate of interest below a certain figure, which may prove important in an era of low interest-rates; namely the intermediate costs of bringing the borrower and the ultimate lender together, and the allowance for risk, especially for moral risk, which the lender requires over and above the pure rate of interest. As the pure rate of interest declines it does not follow that the allowances for expense and risk decline *pari passu*. Thus the rate of interest which the typical borrower has to pay may decline more slowly than the pure rate of interest, and may be incapable of being brought, by the methods of the existing banking and financial organisation, below a certain minimum figure. This is particularly important if the estimation of moral risk is appreciable. For where the risk is due to doubt in the mind of the lender concerning the honesty of the borrower, there is nothing in the mind of a borrower who does not intend to be dishonest to offset the resultant higher charge. It is also important in the case of short-term loans (e.g. bank loans) where the expenses are heavy;—a bank may have to charge its customers $1\frac{1}{2}$ to 2 per cent., even if the pure rate of interest to the lender is nil.

IV

At the cost of anticipating what is more properly the subject of chapter 21 below it may be interesting briefly at this stage to indicate the relationship of the above to the quantity theory of money.

In a static society or in a society in which for any other reason no one feels any uncertainty about the future rates of interest, the liquidity function L_2, or the propensity to hoard (as we might term it), will always be zero in equilibrium. Hence in equilibrium $M_2 = 0$ and $M = M_1$; so that any change in M will cause the rate of interest to fluctuate until income reaches a level at which the change in M_1 is equal to the supposed change in M. Now $M_1 V = Y$, where V is the income-velocity of money as defined above and Y is the aggregate income. Thus if it is practicable to measure the quantity, O, and the price, P, of current output, we have $Y = OP$, and, therefore, $MV = OP$; which is much the same as the quantity theory of money in its traditional form.[1]

For the purposes of the real world it is a great fault in the quantity theory that it does not distinguish between changes in prices which are a function of changes in output, and those which are a function of changes in the wage-unit.[2] The explanation of this omission is, perhaps, to be found in the assumptions that there is no propensity to hoard and that there is always full employment. For in this case, O being constant and M_2 being zero, it follows, if we can take V also as constant, that both the wage-unit and the price-level will be directly proportional to the quantity of money.

[1] If we had defined V, not as equal to Y/M_1 but as equal to Y/M, then, of course, the quantity theory is a truism which holds in all circumstances, though without significance.
[2] This point will be further developed in chapter 21 below.

Chapter 16

SUNDRY OBSERVATIONS ON THE NATURE OF CAPITAL

I

An act of individual saving means—so to speak—a decision not to have dinner to-day. But it does *not* necessitate a decision to have dinner or to buy a pair of boots a week hence or a year hence or to consume any specified thing at any specified date. Thus it depresses the business of preparing to-day's dinner without stimulating the business of making ready for some future act of consumption. It is not a substitution of future consumption-demand for present consumption-demand,—it is a net diminution of such demand. Moreover, the expectation of future consumption is so largely based on current experience of present consumption that a reduction in the latter is likely to depress the former, with the result that the act of saving will not merely depress the price of consumption-goods and leave the marginal efficiency of existing capital unaffected, but may actually tend to depress the latter also. In this event it may reduce present investment-demand as well as present consumption-demand.

If saving consisted not merely in abstaining from present consumption but in placing simultaneously a specific order for future consumption, the effect might indeed be different. For in that case the expectation of some future yield from investment would be improved, and the resources released from preparing for present consumption could be turned over to preparing for the future consumption. Not that they necessarily would be, even in this case, on a scale *equal* to the amount of resources released; since the desired interval of delay might require a method of production so inconveniently 'roundabout' as to have an efficiency well below the current rate of interest, with the result that the favourable effect on employment of the forward order for consumption would eventuate not at once but at some subsequent date, so that the *immediate* effect of the saving would still be adverse to employment. In any case, however, an individual decision to save does not, in actual fact, involve the placing of any specific forward order for consumption, but merely the cancellation of a present order. Thus, since the expectation of consumption is the only *raison d'être* of employment, there should be nothing paradoxical in the conclusion that a diminished propensity to consume has *cet. par.* a depressing effect on employment.

The trouble arises, therefore, because the act of saving implies, not a substitution for present consumption of some specific additional consumption which requires for its preparation just as much immediate economic activity as would have been required by present consumption equal in value to the sum saved, but a desire for 'wealth' as such, that is for a potentiality of consuming an unspecified article at an unspecified time. The absurd, though almost universal, idea that an act of individual saving is just as good for effective demand as an act of individual consumption, has been fostered by the fallacy, much more specious than the conclusion derived from it, that an increased desire to hold wealth, being much the same thing as an increased desire to hold investments, must, by increasing the demand for investments, provide a stimulus to their production; so that current investment is promoted by individual saving to the same extent as present consumption is diminished.

It is of this fallacy that it is most difficult to disabuse men's minds. It comes from believing that the owner of wealth desires a capital-asset *as such*, whereas what he really desires is its *prospective yield*. Now, prospective yield wholly depends on the expectation of future effective demand in relation to future conditions of supply. If, therefore, an act of saving does nothing to improve prospective yield, it does nothing to stimulate investment. Moreover, in order that an individual saver may attain his desired goal of the ownership of wealth, it is not necessary that a *new* capital-asset should be produced wherewith to satisfy him. The mere act of saving by one individual, being *two-sided* as we have shown above, forces some other individual to transfer to him some article of wealth old or new. Every act of saving involves a 'forced' inevitable transfer of wealth to him who saves, though he in his turn may suffer from the saving of others. These transfers of wealth do not require the creation of new wealth—indeed, as we have seen, they may be actively inimical to it. The creation of new wealth wholly depends on the prospective yield of the new wealth reaching the standard set by the current rate of interest. The prospective yield of the marginal new investment is not increased by the fact that someone wishes to increase his wealth, since the prospective yield of the marginal new investment depends on the expectation of a demand for a specific article at a specific date.

Nor do we avoid this conclusion by arguing that what the owner of wealth desires is not a given prospective yield but the best available prospective yield, so that an increased desire to own wealth reduces the prospective yield with which the producers of new investment have to be content. For this overlooks the fact that there is always an alternative to the ownership of real capital-assets, namely the ownership of money and debts; so that the prospective yield with which the producers of new investment have to be content cannot fall below the standard set by the current rate of interest. And the current rate of interest depends, as we have seen, not on the strength of the desire to hold wealth, but on the strengths of the desires to hold it in liquid and in illiquid forms respectively, coupled with the amount of the supply of wealth in the one form relatively to the supply of it in the other. If the reader still finds himself perplexed, let him ask himself why, the quantity of money being unchanged, a fresh act of saving should diminish the sum which it is desired to keep in liquid form at the existing rate of interest.

Certain deeper perplexities, which may arise when we try to probe still further into the whys and wherefores, will be considered in the next chapter.

II

It is much preferable to speak of capital as having a yield over the course of its life in excess of its original cost, than as being *productive*. For the only reason why an asset offers a prospect of yielding during its life services having an aggregate value greater than its initial supply price is because it is *scarce*; and it is kept scarce because of the competition of the rate of interest on money. If capital becomes less scarce, the excess yield will diminish, without its having become less productive—at least in the physical sense.

I sympathise, therefore, with the pre-classical doctrine that everything is *produced* by *labour*, aided by what used to be called art and is now called technique, by natural resources which are free or cost a rent according to their scarcity or abundance, and by the results of past labour, embodied in assets, which also command a price according to their scarcity or abundance. It is preferable to regard labour, including, of course, the personal services of the entrepreneur and his assistants, as the sole factor of production, operating in a given environment of technique, natural resources, capital equipment and effective demand. This partly explains why we have been able to take the unit of labour as the sole physical unit which we require in our economic system, apart from units of money and of time.

It is true that some lengthy or roundabout processes are physically efficient. But so are some short processes. Lengthy processes are not physically efficient because they are long. Some, probably most, lengthy processes would be physically very inefficient, for there are such things as spoiling or wasting with time.[1] With a given labour force there is a definite limit to the quantity of labour embodied in roundabout processes which can be used to advantage. Apart from other considerations, there must be a due proportion between the amount of labour employed in making machines and the amount which will be employed in using them. The ultimate quantity of *value* will not increase indefinitely, relatively to the quantity of labour employed, as the processes adopted become more and more roundabout, even if their physical efficiency is still increasing. Only if the desire to postpone consumption were strong enough to produce a situation in which full employment required a volume of investment so great as to involve a negative marginal efficiency of capital, would a process become advantageous merely because it was lengthy; in which event we should employ physically *inefficient* processes, provided they were sufficiently lengthy for the gain from postponement to outweigh their inefficiency. We should in fact have a situation in which *short* processes would have to be kept sufficiently scarce for their physical efficiency to outweigh the disadvantage of the early delivery of their product. A correct theory, therefore, must be reversible so as to be able to cover the cases of the marginal efficiency of capital corresponding

[1] Cf. Marshall's note on Böhm-Bawerk, *Principles*, p. 583.

either to a positive or to a negative rate of interest; and it is, I think, only the scarcity theory outlined above which is capable of this.

Moreover there are all sorts of reasons why various kinds of services and facilities are scarce and therefore expensive relatively to the quantity of labour involved. For example, smelly processes command a higher reward, because people will not undertake them otherwise. So do risky processes. But we do not devise a productivity theory of smelly or risky processes as such. In short, not all labour is accomplished in equally agreeable attendant circumstances; and conditions of equilibrium require that articles produced in less agreeable attendant circumstances (characterised by smelliness, risk or the lapse of time) must be kept sufficiently scarce to command a higher price. But if the lapse of time becomes an agreeable attendant circumstance, which is a quite possible case and already holds for many individuals, then, as I have said above, it is the short processes which must be kept sufficiently scarce.

Given the optimum amount of roundaboutness, we shall, of course, select the most efficient roundabout processes which we can find up to the required aggregate. But the optimum amount itself should be such as to provide at the appropriate dates for that part of consumers' demand which it is desired to defer. In optimum conditions, that is to say, production should be so organised as to produce in the most efficient manner compatible with delivery at the dates at which consumers' demand is expected to become effective. It is no use to produce for delivery at a different date from this, even though the physical output could be increased by changing the date of delivery;— except in so far as the prospect of a larger meal, so to speak, induces the consumer to anticipate or postpone the hour of dinner. If, after hearing full particulars of the meals he can get by fixing dinner at different hours, the consumer is expected to decide in favour of 8 o'clock, it is the business of the cook to provide the best dinner he can for service at that hour, irrespective of whether 7.30, 8 o'clock or 8.30 is the hour which would suit him best if time counted for nothing, one way or the other, and his only task was to produce the absolutely best dinner. In some phases of society it may be that we could get physically better dinners by dining later than we do; but it is equally conceivable in other phases that we could get better dinners by dining earlier. Our theory must, as I have said above, be applicable to both contingencies.

If the rate of interest were zero, there would be an optimum interval for any given article between the average date of input and the date of consumption, for which labour cost would be a minimum;—a shorter process of production would be less efficient technically, whilst a longer process would also be less efficient by reason of storage costs and deterioration. If, however, the rate of interest exceeds zero, a new element of cost is introduced which increases with the length of the process, so that the optimum interval will be shortened, and the current input to provide for the eventual delivery of the article will have to be curtailed until the prospective price has increased sufficiently to cover the increased cost—a cost which will be increased both by the interest charges and also by the diminished efficiency of the shorter method of production. Whilst if the rate of interest falls below zero (assuming this to be technically possible), the opposite is the case. Given the prospective consumers' demand, current input to-day has to compete, so to speak, with the alternative of starting input at a later date; and, consequently, current input will only be worth while when the greater cheapness, by reason of greater technical efficiency or prospective price changes, of producing later on rather than now, is insufficient to offset the smaller return from negative interest. In the case of the great majority of articles it would involve great technical *in*efficiency to start up their input more than a very modest length of time ahead of their prospective consumption. Thus even if the rate of interest is zero, there is a strict limit to the proportion of prospective consumers' demand which it is profitable to begin providing for in advance; and, as the rate of interest rises, the proportion of the prospective consumers' demand for which it pays to produce to-day shrinks *pari passu*.

III

We have seen that capital has to be kept scarce enough in the long-period to have a marginal efficiency which is at least equal to the rate of interest for a period equal to the life of the capital, as determined by psychological and institutional conditions. What would this involve for a society which finds itself so well equipped with capital that its marginal efficiency is zero and would be negative with any additional investment; yet possessing a monetary system, such that money will 'keep' and involves negligible costs of storage and safe custody, with the result that in practice interest cannot be negative; and, in conditions of full employment, disposed to save?

If, in such circumstances, we start from a position of full employment, entrepreneurs will necessarily make losses if they continue to offer employment on a scale which will utilise the whole of the existing stock of capital. Hence the stock of capital and the level of employment will have to shrink until the community becomes so impoverished that the aggregate of saving has become zero, the positive saving of some individuals or groups being offset by the negative saving of others. Thus for a society such as we have supposed, the position of equilibrium, under conditions of *laissez-faire*, will be one in which employment is low enough and the standard of life sufficiently miserable to bring savings to zero. More probably there will be a cyclical movement round this equilibrium position. For if there is still room for uncertainty about the future, the marginal efficiency of capital will occasionally rise above zero leading to a 'boom', and in the succeeding 'slump' the stock of capital may fall for a time below the level which will yield a marginal efficiency of zero in the long run. Assuming correct foresight, the equilibrium stock of capital which will have a marginal efficiency of precisely zero will, of course, be a smaller stock than would correspond to full employment of the available labour; for it will be the equipment which corresponds to that proportion of unemployment which ensures zero saving.

The only alternative position of equilibrium would be given by a situation in which a stock of capital sufficiently great to have a marginal efficiency of zero also represents an amount of wealth sufficiently great to satiate to the full the aggregate desire on the part of the public to make provision for the future, even with full employment, in circumstances where no bonus is obtainable in the form of interest. It would, however, be an unlikely coincidence that the propensity to save in conditions of full employment should become satisfied just at the point where the stock of capital reaches the level where its marginal efficiency is zero. If, therefore, this more favourable possibility comes to the rescue, it will probably take effect, not just at the point where the rate of interest is vanishing, but at some previous point during the gradual decline of the rate of interest.

We have assumed so far an institutional factor which prevents the rate of interest from being negative, in the shape of money which has negligible carrying costs. In fact, however, institutional and psychological factors are present which set a limit much above zero to the practicable decline in the rate of interest. In particular the costs of bringing borrowers and lenders together and uncertainty as to the future of the rate of interest, which we have examined above, set a lower limit, which in present circumstances may perhaps be as high as 2 or $2\frac{1}{2}$ per cent on long term. If this should prove correct, the awkward possibilities of an increasing stock of wealth, in conditions where the rate of interest can fall no further under *laissez-faire*, may soon be realised in actual experience. Moreover if the minimum level to which it is practicable to bring the rate of interest is appreciably above zero, there is less likelihood of the aggregate desire to accumulate wealth being satiated before the rate of interest has reached its minimum level.

The post-war experiences of Great Britain and the United States are, indeed, actual examples of how an accumulation of wealth, so large that its marginal efficiency has fallen more rapidly than the rate of interest can fall in the face of the prevailing institutional and psychological factors, can interfere, in conditions mainly of *laissez-faire*, with a reasonable level of employment and with the standard of life which the technical conditions of production are capable of furnishing.

It follows that of two equal communities, having the same technique but different stocks of capital, the community with the smaller stocks of capital may be able for the time being to enjoy a higher standard of life than the community with the larger stock; though when the poorer community has caught up the rich— as, presumably, it eventually will—then both alike will suffer the fate of Midas. This disturbing conclusion depends, of course, on the assumption that the propensity to consume and the rate of investment are not deliberately controlled in the social interest but are mainly left to the influences of *laissez-faire*.

If—for whatever reason—the rate of interest cannot fall as fast as the marginal efficiency of capital would fall with a rate of accumulation corresponding to what the community would choose to save at a rate of interest equal to the marginal efficiency of capital in conditions of full employment, then even a diversion of the desire to hold wealth towards assets, which will in fact yield no economic fruits whatever, will increase economic well-being. In so far as millionaires find their satisfaction in building mighty mansions to contain their bodies when alive and pyramids to shelter them after death, or, repenting of their sins, erect cathedrals and endow monasteries or foreign missions, the day when abundance of capital will interfere with abundance of output may be postponed. 'To dig holes in the ground', paid for out of savings, will increase, not only employment, but the real national dividend of useful goods and services. It is not reasonable, however, that a sensible community should be content to remain dependent on such fortuitous and often wasteful mitigations when once we understand the influences upon which effective demand depends.

IV

Let us assume that steps are taken to ensure that the rate of interest is consistent with the rate of investment which corresponds to full employment. Let us assume, further, that State action enters in as a balancing factor to provide that the growth of capital equipment shall be such as to approach saturation-point at a rate which does not put a disproportionate burden on the standard of life of the present generation.

On such assumptions I should guess that a properly run community equipped with modern technical resources, of which the population is not increasing rapidly, ought to be able to bring down the marginal efficiency of capital in equilibrium approximately to zero within a single generation; so that we should attain the conditions of a quasi-stationary community where change and progress would result only from changes in technique, taste, population and institutions, with the products of capital selling at a price pro- portioned to the labour, etc., embodied in them on just the same principles as govern the prices of con- sumption-goods into which capital-charges enter in an insignificant degree.

If I am right in supposing it to be comparatively easy to make capital-goods so abundant that the marginal efficiency of capital is zero, this may be the most sensible way of gradually getting rid of many of the objectionable features of capitalism. For a little reflection will show what enormous social changes would result from a gradual disappearance of a rate of return on accumulated wealth. A man would still be free to accumulate his earned income with a view to spending it at a later date. But his accumulation would not grow. He would simply be in the position of Pope's father, who, when he retired from business, carried a chest of guineas with him to his villa at Twickenham and met his household expenses from it as required.

Though the rentier would disappear, there would still be room, nevertheless, for enterprise and skill in the estimation of prospective yields about which opinions could differ. For the above relates primarily to the pure rate of interest apart from any allowance for risk and the like, and not to the gross yield of assets including the return in respect of risk. Thus unless the pure rate of interest were to be held at a negative figure, there would still be a positive yield to skilled investment in individual assets having a doubtful pro- spective yield. Provided there was some measurable unwillingness to undertake risk, there would also be a positive net yield from the aggregate of such assets over a period of time. But it is not unlikely that, in such circumstances, the eagerness to obtain a yield from doubtful investments might be such that they would show in the aggregate a *negative* net yield.

Chapter 17

THE ESSENTIAL PROPERTIES OF
INTEREST AND MONEY

I

It seems, then, that the *rate of interest on money* plays a
peculiar part in setting a limit to the level of employment, since it sets a standard to which the marginal
efficiency of a capital-asset must attain if it is to be newly
produced. That this should be so, is, at first sight, most
perplexing. It is natural to enquire wherein the
peculiarity of money lies as distinct from other assets,
whether it is only money which has a rate of interest,
and what would happen in a non-monetary economy.
Until we have answered these questions, the full
significance of our theory will not be clear.

The money-rate of interest—we may remind the
reader—is nothing more than the percentage excess of
a sum of money contracted for forward delivery, e.g. a
year hence, over what we may call the 'spot' or cash
price of the sum thus contracted for forward delivery.
It would seem, therefore, that for every kind of capital-
asset there must be an analogue of the rate of interest
on money. For there is a definite quantity of (e.g.)
wheat to be delivered a year hence which has the same
exchange value to-day as 100 quarters of wheat for
'spot' delivery. If the former quantity is 105 quarters,
we may say that the wheat-rate of interest is 5 per cent
per annum; and if it is 95 quarters, that it is *minus*
5 per cent per annum. Thus for every durable commodity we have a rate of interest in terms of itself,—a

wheat-rate of interest, a copper-rate of interest, a house-rate of interest, even a steel-plant-rate of interest.

The difference between the 'future' and 'spot' contracts for a commodity, such as wheat, which are quoted in the market, bears a definite relation to the wheat-rate of interest, but, since the future contract is quoted in terms of money for forward delivery and not in terms of wheat for spot delivery, it also brings in the money-rate of interest. The exact relationship is as follows:

Let us suppose that the spot price of wheat is £100 per 100 quarters, that the price of the 'future' contract for wheat for delivery a year hence is £107 per 100 quarters, and that the money-rate of interest is 5 per cent; what is the wheat-rate of interest? £100 spot will buy £105 for forward delivery, and £105 for forward delivery will buy $\frac{105}{107}.100$ (= 98) quarters for forward delivery. Alternatively £100 spot will buy 100 quarters of wheat for spot delivery. Thus 100 quarters of wheat for spot delivery will buy 98 quarters for forward delivery. It follows that the wheat-rate of interest is *minus* 2 per cent per annum.[1]

It follows from this that there is no reason why their rates of interest should be the same for different commodities,—why the wheat-rate of interest should be equal to the copper-rate of interest. For the relation between the 'spot' and 'future' contracts, as quoted in the market, is notoriously different for different commodities. This, we shall find, will lead us to the clue we are seeking. For it may be that it is the *greatest* of the own-rates of interest (as we may call them) which rules the roost (because it is the greatest of these rates that the marginal efficiency of a capital-asset must attain if it is to be newly produced); and that there are reasons why it is the money-rate of interest which is often the greatest (because, as we shall find, certain

---

[1] This relationship was first pointed out by Mr Sraffa, *Economic Journal*, March 1932, p. 50.

forces, which operate to reduce the own-rates of interest of other assets, do not operate in the case of money).

It may be added that, just as there are differing commodity-rates of interest at any time, so also exchange dealers are familiar with the fact that the rate of interest is not even the same in terms of two different moneys, e.g. sterling and dollars. For here also the difference between the 'spot' and 'future' contracts for a foreign money in terms of sterling are not, as a rule, the same for different foreign moneys.

Now each of these commodity standards offers us the same facility as money for measuring the marginal efficiency of capital. For we can take any commodity we choose, e.g. wheat; calculate the wheat-value of the prospective yields of any capital asset; and the rate of discount which makes the present value of this series of wheat annuities equal to the present supply price of the asset in terms of wheat gives us the marginal efficiency of the asset in terms of wheat. If no change is expected in the relative value of two alternative standards, then the marginal efficiency of a capital-asset will be the same in whichever of the two standards it is measured, since the numerator and denominator of the fraction which leads up to the marginal efficiency will be changed in the same proportion. If, however, one of the alternative standards is expected to change in value in terms of the other, the marginal efficiencies of capital-assets will be changed by the same percentage, according to which standard they are measured in. To illustrate this let us take the simplest case where wheat, one of the alternative standards, is expected to appreciate at a steady rate of a per cent per annum in terms of money; the marginal efficiency of an asset, which is x per cent in terms of money, will then be $x - a$ per cent in terms of wheat. Since the marginal efficiencies of all capital-assets will be altered by the same amount, it follows that their order of magnitude will be the same irrespective of the standard which is selected.

If there were some composite commodity which could be regarded strictly speaking as representative, we could regard the rate of interest and the marginal efficiency of capital in terms of this commodity as being, in a sense, uniquely *the* rate of interest and *the* marginal efficiency of capital. But there are, of course, the same obstacles in the way of this as there are to setting up a unique standard of value.

So far, therefore, the money-rate of interest has no uniqueness compared with other rates of interest, but is on precisely the same footing. Wherein, then, lies the peculiarity of the money-rate of interest which gives it the predominating practical importance attributed to it in the preceding chapters? Why should the volume of output and employment be more intimately bound up with the money-rate of interest than with the wheat-rate of interest or the house-rate of interest?

II

Let us consider what the various commodity-rates of interest over a period of (say) a year are likely to be for different types of assets. Since we are taking each commodity in turn as the standard, the returns on each commodity must be reckoned in this context as being measured in terms of itself.

There are three attributes which different types of assets possess in different degrees; namely, as follows:

(i) Some assets produce a yield or output q, measured in terms of themselves, by assisting some process of production or supplying services to a consumer.

(ii) Most assets, except money, suffer some wastage or involve some cost through the mere passage of time (apart from any change in their relative value), irrespective of their being used to produce a yield; i.e. they involve a carrying cost c measured in terms of themselves. It does not matter for our present purpose exactly where we draw the line between the costs which we deduct before calculating q and those which we include in c, since in what follows we shall be exclusively concerned with $q - c$.

(iii) Finally, the power of disposal over an asset during a period may offer a potential convenience or security, which is not equal for assets of different kinds, though the assets themselves are of equal initial value. There is, so to speak, nothing to show for this at the end of the period in the shape of output; yet it is something for which people are ready to pay something. The amount (measured in terms of itself) which they are willing to pay for the potential convenience or security given by this power of disposal (exclusive of yield or carrying cost attaching to the asset), we shall call its liquidity-premium l.

It follows that the total return expected from the ownership of an asset over a period is equal to its yield *minus* its carrying cost *plus* its liquidity-premium, i.e. to $q - c + l$. That is to say, $q - c + l$ is the own-rate of interest of any commodity, where q, c and l are measured in terms of itself as the standard.

It is characteristic of instrumental capital (e.g. a machine) or of consumption capital (e.g. a house) which is in use, that its yield should normally exceed its carrying cost, whilst its liquidity-premium is probably negligible; of a stock of liquid goods or of surplus laid-up instrumental or consumption capital that it should incur a carrying cost in terms of itself without any yield to set off against it, the liquidity-premium in this case also being usually negligible as soon as stocks exceed a moderate level, though capable of being significant in special circumstances; and of money that its yield is *nil*, and its carrying cost negligible, but its liquidity-premium substantial. Different commodities may, indeed, have differing degrees of liquidity-premium amongst themselves, and money may incur some degree of carrying costs, e.g. for safe custody. But it is an essential difference between money and all (or most) other assets that in the case of money its liquidity-premium much exceeds its carrying cost, whereas in the case of other assets their carrying cost much exceeds their liquidity-premium. Let us, for purposes of illustration, assume that on houses the yield is q_1 and the carrying cost and liquidity-premium negligible; that on wheat the carrying cost is c_2 and the yield and liquidity-premium negligible; and that on money the liquidity-premium is l_3 and the yield and carrying cost negligible. That is to say, q_1 is the house-rate of interest, $-c_2$ the wheat-rate of interest, and l_3 the money-rate of interest.

To determine the relationships between the expected returns on different types of assets which are consistent with equilibrium, we must also know what the changes in relative values during the year are expected to be. Taking money (which need only be a money of account for this purpose, and we could equally well take wheat) as our standard of measurement, let the expected percentage appreciation (or depreciation) of houses be a_1 and of wheat a_2. q_1, $-c_2$ and l_3 we have called the own-rates of interest of houses, wheat and money in terms of themselves as the standard of value; i.e. q_1 is the house-rate of interest in terms of houses, $-c_2$ is the wheat-rate of interest in terms of wheat, and l_3 is the money-rate of interest in terms of money. It will also be useful to call $a_1 + q_1$, $a_2 - c_2$ and l_3, which stand for the same quantities reduced to money as the standard of value, the house-rate of money-interest, the wheat-rate of money-interest and the money-rate of money-interest respectively. With this notation it is easy to see that the demand of wealth-owners will be directed to houses, to wheat or to money, according as $a_1 + q_1$ or $a_2 - c_2$ or l_3 is greatest. Thus in equilibrium the demand-prices of houses and wheat in terms of money will be such that there is nothing to choose in the way of advantage between the alternatives;—i.e. $a_1 + q_1$, $a_2 - c_2$ and l_3 will be *equal*. The choice of the standard of value will make no difference to this result because a shift from one standard to another will change all the terms equally, i.e. by an amount equal to the expected rate of appreciation (or depreciation) of the new standard in terms of the old.

Now those assets of which the normal supply-price is less than the demand-price will be newly produced; and these will be those assets of which the marginal efficiency would be greater (on the basis of their normal supply-price) than the rate of interest (both being measured in the same standard of value whatever it is). As the stock of the assets, which begin by having a marginal efficiency at least equal to the rate of interest, is increased, their marginal efficiency (for reasons, sufficiently obvious, already given) tends to fall. Thus a point will come at which it no longer pays to produce them, *unless the rate of interest falls* pari passu. When there is *no* asset of which the marginal efficiency reaches the rate of interest, the further production of capital-assets will come to a standstill.

Let us suppose (as a mere hypothesis at this stage of the argument) that there is some asset (e.g. money) of which the rate of interest is fixed (or declines more slowly as output increases than does any other commodity's rate of interest); how is the position adjusted? Since $a_1 + q_1$, $a_2 - c_2$ and l_3 are necessarily equal, and since l_3 by hypothesis is either fixed or falling more slowly than q_1 or $-c_2$, it follows that a_1 and a_2 must be rising. In other words, the present money-price of every commodity other than money tends to fall relatively to its expected future price. Hence, if q_1 and $-c_2$ continue to fall, a point comes at which it is not profitable to produce any of the commodities, unless the cost of production at some future date is expected to rise above the present cost by an amount which will cover the cost of carrying a stock produced now to the date of the prospective higher price.

It is now apparent that our previous statement to the effect that it is the money-rate of interest which sets a limit to the rate of output, is not strictly correct. We should have said that it is that asset's rate of interest which declines most slowly as the stock of assets in general increases, which eventually knocks out the profitable production of each of the others,—except in the contingency, just mentioned, of a special relationship between the present and prospective costs of production. As output increases, own-rates of interest decline to levels at which one asset after another falls below the standard of profitable production;—until, finally, one or more own-rates of interest remain at a level which is above that of the marginal efficiency of any asset whatever.

If by *money* we mean the standard of value, it is clear that it is not necessarily the money-rate of interest which makes the trouble. We could not get out of our difficulties (as some have supposed) merely by decreeing that wheat or houses shall be the standard of value instead of gold or sterling. For, it now appears that the same difficulties will ensue if there continues to exist *any* asset of which the own-rate of interest is reluctant to decline as output increases. It may be, for example, that gold will continue to fill this rôle in a country which has gone over to an inconvertible paper standard.

III

In attributing, therefore, a peculiar significance to the money-rate of interest, we have been tacitly assuming that the kind of money to which we are accustomed has some special characteristics which lead to its own-rate of interest in terms of itself as standard being more reluctant to fall as the stock of assets in general increases than the own-rates of interest of any other assets in terms of themselves. Is this assumption justified? Reflection shows, I think, that the following peculiarities, which commonly characterise money as we know it, are capable of justifying it. To the extent that the established standard of value has these peculiarities, the summary statement, that it is the money-rate of interest which is the significant rate of interest, will hold good.

(i) The first characteristic which tends towards the above conclusion is the fact that money has, both in the long and in the short period, a zero, or at any rate a very small, elasticity of production, so far as the power of private enterprise is concerned, as distinct from the monetary authority;—elasticity of production[1] meaning, in this context, the response of the quantity of labour applied to producing it to a rise in the quantity of labour which a unit of it will command. Money, that is to say, cannot be readily produced;—labour cannot be turned on at will by entrepreneurs to produce money in increasing quantities as its price rises in terms of the wage-unit. In the case of an inconvertible managed currency this condition is strictly satisfied. But in the case of a gold-standard currency it is also approximately so, in the sense that the maximum proportional addition to the quantity of labour which can be thus employed is very small, except indeed in a country of which gold-mining is the major industry.

Now, in the case of assets having an elasticity of production, the reason why we assumed their own-rate of interest to decline was because we assumed the stock of them to increase as the result of a higher rate of output. In the case of money, however—postponing, for the moment, our consideration of the effects of reducing the wage-unit or of a deliberate increase in its supply by the monetary authority—the supply is fixed. Thus the characteristic that money cannot be readily produced by labour gives at once some *prima facie* presumption for the view that its own-rate of interest will be relatively reluctant to fall; whereas if money could be grown like a crop or manufactured

---

[1] See chapter 20.

like a motor-car, depressions would be avoided or mitigated because, if the price of other assets was tending to fall in terms of money, more labour would be diverted into the production of money;—as we see to be the case in gold-mining countries, though for the world as a whole the maximum diversion in this way is almost negligible.

(ii) Obviously, however, the above condition is satisfied, not only by money, but by all pure rent-factors, the production of which is completely inelastic. A second condition, therefore, is required to distinguish money from other rent elements.

The second *differentia* of money is that it has an elasticity of substitution equal, or nearly equal, to zero; which means that as the exchange value of money rises there is no tendency to substitute some other factor for it;—except, perhaps, to some trifling extent, where the money-commodity is also used in manufacture or the arts. This follows from the peculiarity of money that its utility is solely derived from its exchange-value, so that the two rise and fall *pari passu*, with the result that as the exchange value of money rises there is no motive or tendency, as in the case of rent-factors, to substitute some other factor for it.

Thus, not only is it impossible to turn more labour on to producing money when its labour-price rises, but money is a bottomless sink for purchasing power, when the demand for it increases, since there is no value for it at which demand is diverted—as in the case of other rent-factors—so as to slop over into a demand for other things.

The only qualification to this arises when the rise in the value of money leads to uncertainty as to the future maintenance of this rise; in which event, a_1 and a_2 are increased, which is tantamount to an increase in the commodity-rates of money-interest and is, therefore, stimulating to the output of other assets.

(iii) Thirdly, we must consider whether these conclusions are upset by the fact that, even though the quantity of money cannot be increased by diverting labour into producing it, nevertheless an assumption that its effective supply is rigidly fixed would be inaccurate. In particular, a reduction of the wage-unit will release cash from its other uses for the satisfaction of the liquidity-motive; whilst, in addition to this, as money-values fall, the stock of money will bear a higher proportion to the total wealth of the community.

It is not possible to dispute on purely theoretical grounds that this reaction might be capable of allowing an adequate decline in the money-rate of interest. There are, however, several reasons, which taken in combination are of compelling force, why in an economy of the type to which we are accustomed it is very probable that the money-rate of interest will often prove reluctant to decline adequately:

(a) We have to allow, first of all, for the reactions of a fall in the wage-unit on the marginal efficiencies of other assets in terms of money;—for it is the *difference* between these and the money-rate of interest with which we are concerned. If the effect of the fall in the wage-unit is to produce an expectation that it will subsequently rise again, the result will be wholly favourable. If, on the contrary, the effect is to produce an expectation of a further fall, the reaction on the marginal efficiency of capital may offset the decline in the rate of interest.[1]

(b) The fact that wages tend to be sticky in terms of money, the money-wage being more stable than the real wage, tends to limit the readiness of the wage-unit to fall in terms of money. Moreover, if this were not so, the position might be worse rather than better; because, if money-wages were to fall easily, this might often tend to create an expectation of a further fall with unfavourable reactions on the marginal efficiency of capital.

[1] This is a matter which will be examined in greater detail in chapter 19 below.

Furthermore, if wages were to be fixed in terms of some other commodity, e.g. wheat, it is improbable that they would continue to be sticky. It is because of money's other characteristics—those, especially, which make it *liquid*—that wages, when fixed in terms of it, tend to be sticky.[1]

(*c*) Thirdly, we come to what is the most fundamental consideration in this context, namely, the characteristics of money which satisfy liquidity-preference. For, in certain circumstances such as will often occur, these will cause the rate of interest to be insensitive, particularly below a certain figure,[2] even to a substantial increase in the quantity of money in proportion to other forms of wealth. In other words, beyond a certain point money's yield from liquidity does not fall in response to an increase in its quantity to anything approaching the extent to which the yield from other types of assets falls when their quantity is comparably increased.

In this connection the low (or negligible) carrying-costs of money play an essential part. For if its carrying costs were material, they would offset the effect of expectations as to the prospective value of money at future dates. The readiness of the public to increase their stock of money in response to a comparatively small stimulus is due to the advantages of liquidity (real or supposed) having no offset to contend with in the shape of carrying-costs mounting steeply with the lapse of time. In the case of a commodity other than money a modest stock of it may offer some convenience to users of the commodity. But even though a larger stock might have some attractions as representing a store of wealth of stable value, this would be offset by its carrying-costs in the shape of storage, wastage, etc.

---

[1] If wages (and contracts) were fixed in terms of wheat, it might be that wheat would acquire some of money's liquidity premium;—we will return to this question in (IV) below.

[2] See p. 172 above.

Hence, after a certain point is reached, there is necessarily a loss in holding a greater stock.

In the case of money, however, this, as we have seen, is not so,—and for a variety of reasons, namely, those which constitute money as being, in the estimation of the public, *par excellence* 'liquid'. Thus those reformers, who look for a remedy by creating artificial carrying-costs for money through the device of requiring legal-tender currency to be periodically stamped at a prescribed cost in order to retain its quality as money, or in analogous ways, have been on the right track; and the practical value of their proposals deserves consideration.

The significance of the money-rate of interest arises, therefore, out of the combination of the characteristics that, through the working of the liquidity-motive, this rate of interest may be somewhat unresponsive to a change in the proportion which the quantity of money bears to other forms of wealth measured in money, and that money has (or may have) zero (or negligible) elasticities both of production and of substitution. The first condition means that demand may be predominantly directed to money, the second that when this occurs labour cannot be employed in producing more money, and the third that there is no mitigation at any point through some other factor being capable, if it is sufficiently cheap, of doing money's duty equally well. The only relief—apart from changes in the marginal efficiency of capital—can come (so long as the propensity towards liquidity is unchanged) from an increase in the quantity of money, or—which is formally the same thing—a rise in the value of money which enables a given quantity to provide increased money-services.

Thus a rise in the money-rate of interest retards the output of all the objects of which the production is elastic without being capable of stimulating the output of money (the production of which is, by hypothesis,

234

perfectly inelastic). The money-rate of interest, by setting the pace for all the other commodity-rates of interest, holds back investment in the production of these other commodities without being capable of stimulating investment for the production of money, which by hypothesis cannot be produced. Moreover, owing to the elasticity of demand for liquid cash in terms of debts, a small change in the conditions governing this demand may not much alter the money-rate of interest, whilst (apart from official action) it is also impracticable, owing to the inelasticity of the production of money, for natural forces to bring the money-rate of interest down by affecting the supply side. In the case of an ordinary commodity, the inelasticity of the demand for liquid stocks of it would enable small changes on the demand side to bring its rate of interest up or down with a rush, whilst the elasticity of its supply would also tend to prevent a high premium on spot over forward delivery. Thus with other commodities left to themselves, 'natural forces,' i.e. the ordinary forces of the market, would tend to bring their rate of interest down until the emergence of full employment had brought about for commodities generally the inelasticity of supply which we have postulated as a normal characteristic of money. Thus in the absence of money and in the absence—we must, of course, also suppose—of any other commodity with the assumed characteristics of money, the rates of interest would only reach equilibrium when there is full employment.

Unemployment develops, that is to say, because people want the moon;—men cannot be employed when the object of desire (i.e. money) is something which cannot be produced and the demand for which cannot be readily choked off. There is no remedy but to persuade the public that green cheese is practically the same thing and to have a green cheese factory (i.e. a central bank) under public control.

It is interesting to notice that the characteristic

235

which has been traditionally supposed to render gold especially suitable for use as the standard of value, namely, its inelasticity of supply, turns out to be precisely the characteristic which is at the bottom of the trouble.

Our conclusion can be stated in the most general form (taking the propensity to consume as given) as follows. No further increase in the rate of investment is possible when the greatest amongst the own-rates of own-interest of all available assets is equal to the greatest amongst the marginal efficiencies of all assets, measured in terms of the asset whose own-rate of own-interest is greatest.

In a position of full employment this condition is necessarily satisfied. But it may also be satisfied before full employment is reached, if there exists some asset, having zero (or relatively small) elasticities of production and substitution,[1] whose rate of interest declines more closely, as output increases, than the marginal efficiencies of capital-assets measured in terms of it.

IV

We have shown above that for a commodity to be the standard of value is not a sufficient condition for that commodity's rate of interest to be the significant rate of interest. It is, however, interesting to consider how far those characteristics of money as we know it, which make the money-rate of interest the significant rate, are bound up with money being the standard in which debts and wages are usually fixed. The matter requires consideration under two aspects.

In the first place, the fact that contracts are fixed, and wages are usually somewhat stable, in terms of money unquestionably plays a large part in attracting to money so high a liquidity-premium. The conveni-

---

[1] A *zero* elasticity is a more stringent condition than is necessarily required.

ence of holding assets in the same standard as that in which future liabilities may fall due and in a standard in terms of which the future cost of living is expected to be relatively stable, is obvious. At the same time the expectation of relative stability in the future money-cost of output might not be entertained with much confidence if the standard of value were a commodity with a high elasticity of production. Moreover, the low carrying-costs of money as we know it play quite as large a part as a high liquidity-premium in making the money-rate of interest the significant rate. For what matters is the *difference* between the liquidity-premium and the carrying-costs; and in the case of most commodities, other than such assets as gold and silver and bank-notes, the carrying-costs are at least as high as the liquidity-premium ordinarily attaching to the standard in which contracts and wages are fixed, so that, even if the liquidity-premium now attaching to (e.g.) sterling-money were to be transferred to (e.g.) wheat, the wheat-rate of interest would still be unlikely to rise above zero. It remains the case, therefore, that, whilst the fact of contracts and wages being fixed in terms of money considerably enhances the significance of the money-rate of interest, this circumstance is, nevertheless, probably insufficient by itself to produce the observed characteristics of the money-rate of interest.

The second point to be considered is more subtle. The normal expectation that the value of output will be more stable in terms of money than in terms of any other commodity, depends of course, not on wages being arranged in terms of money, but on wages being relatively *sticky* in terms of money. What, then, would the position be if wages were expected to be more sticky (i.e. more stable) in terms of some one or more commodities other than money, than in terms of money itself? Such an expectation requires, not only that the costs of the commodity in question are expected to be relatively constant in terms of the wage-unit for a greater or smaller scale of output both in the short and in the long period, but also that any surplus over the current demand at cost-price can be taken into stock without cost, i.e. that its liquidity-premium exceeds its carrying-costs (for, otherwise, since there is no hope of profit from a higher price, the carrying of a stock must necessarily involve a loss). If a commodity can be found to satisfy these conditions, then, assuredly, it might be set up as a rival to money. Thus it is not logically impossible that there should be a commodity in terms of which the value of output is expected to be more stable than in terms of money. But it does not seem probable that any such commodity exists.

I conclude, therefore, that the commodity, in terms of which wages are expected to be most sticky, cannot be one whose elasticity of production is not least, and for which the excess of carrying-costs over liquidity-premium is not least. In other words, the expectation of a relative stickiness of wages in terms of money is a corollary of the excess of liquidity-premium over carrying-costs being greater for money than for any other asset.

Thus we see that the various characteristics, which combine to make the money-rate of interest significant, interact with one another in a cumulative fashion. The fact that money has low elasticities of production and substitution and low carrying-costs tends to raise the expectation that money-wages will be relatively stable; and this expectation enhances money's liquidity-premium and prevents the exceptional correlation between the money-rate of interest and the marginal efficiencies of other assets which might, if it could exist, rob the money-rate of interest of its sting.

Professor Pigou (with others) has been accustomed to assume that there is a presumption in favour of real wages being more stable than money-wages. But this could only be the case if there were a presumption in favour of stability of employment. Moreover, there is also the difficulty that wage-goods have a high carrying-cost. If, indeed, some attempt were made to stabilise real wages by fixing wages in terms of wage-goods, the effect could only be to cause a violent oscillation of money-prices. For every small fluctuation in the propensity to consume and the inducement to invest would cause money-prices to rush violently between zero and infinity. That money-wages should be more stable than real wages is a condition of the system possessing inherent stability.

Thus the attribution of relative stability to real wages is not merely a mistake in fact and experience. It is also a mistake in logic, if we are supposing that the system in view is stable, in the sense that small changes in the propensity to consume and the inducement to invest do not produce violent effects on prices.

V

As a footnote to the above, it may be worth emphasising what has been already stated above, namely, that 'liquidity' and 'carrying-costs' are both a matter of degree; and that it is only in having the former high relatively to the latter that the peculiarity of 'money' consists.

Consider, for example, an economy in which there is no asset for which the liquidity-premium is always in excess of the .carrying-costs; which is the best definition I can give of a so-called 'non-monetary' economy. There exists nothing, that is to say, but particular consumables and particular capital equipments more or less differentiated according to the character of the consumables which they can yield up, or assist to yield up, over a greater or a shorter period of time; all of which, unlike cash, deteriorate or involve expense, if they are kept in stock, to a value in excess of any liquidity-premium which may attach to them.

18-2

In such an economy capital equipments will differ from one another (*a*) in the variety of the consumables in the production of which they are capable of assisting, (*b*) in the stability of value of their output (in the sense in which the value of bread is more stable through time than the value of fashionable novelties), and (*c*) in the rapidity with which the wealth embodied in them can become 'liquid', in the sense of producing output, the proceeds of which can be re-embodied if desired in quite a different form.

The owners of wealth will then weigh the lack of 'liquidity' of different capital equipments in the above sense as a medium in which to hold wealth against the best available actuarial estimate of their prospective yields after allowing for risk. The liquidity-premium, it will be observed, is partly similar to the risk-premium, but partly different;—the difference corresponding to the difference between the best estimates we can make of probabilities and the confidence with which we make them.[1] When we were dealing, in earlier chapters, with the estimation of prospective yield, we did not enter into detail as to how the estimation is made: and to avoid complicating the argument, we did not distinguish differences in liquidity from differences in risk proper. It is evident, however, that in calculating the own-rate of interest we must allow for both.

There is, clearly, no absolute standard of 'liquidity' but merely a scale of liquidity—a varying premium of which account has to be taken, in addition to the yield of use and the carrying-costs, in estimating the comparative attractions of holding different forms of wealth. The conception of what contributes to 'liquidity' is a partly vague one, changing from time to time and depending on social practices and institutions. The order of preference in the minds of owners of wealth in which at any given time they express their feelings about liquidity is, however, definite and is all we require

[1] Cf. the footnote to p. 148 above.

for our analysis of the behaviour of the economic system.

It may be that in certain historic environments the possession of land has been characterised by a high liquidity-premium in the minds of owners of wealth; and since land resembles money in that its elasticities of production and substitution may be very low,[1] it is conceivable that there have been occasions in history in which the desire to hold land has played the same rôle in keeping up the rate of interest at too high a level which money has played in recent times. It is difficult to trace this influence quantitatively owing to the absence of a forward price for land in terms of itself which is strictly comparable with the rate of interest on a money debt. We have, however, something which has, at times, been closely analogous, in the shape of high rates of interest on mortgages.[2] The high rates of interest from mortgages on land, often exceeding the probable net yield from cultivating the land, have been a familiar feature of many agricultural economies. Usury laws have been directed primarily against encumbrances of this character. And rightly so. For in earlier social organisation where long-term bonds in the modern sense were non-existent, the competition of a high interest-rate on mortgages may well have had the same effect in retarding the growth of wealth from current investment in newly produced capital-assets, as high interest rates on long-term debts have had in more recent times.

[1] The attribute of 'liquidity' is by no means independent of the presence of these two characteristics. For it is unlikely that an asset, of which the supply can be easily increased or the desire for which can be easily diverted by a change in relative price, will possess the attribute of 'liquidity' in the minds of owners of wealth. Money itself rapidly loses the attribute of 'liquidity' if its future supply is expected to undergo sharp changes.

[2] A mortgage and the interest thereon are, indeed, fixed in terms of money. But the fact that the mortgagor has the option to deliver the land itself in discharge of the debt—and must so deliver it if he cannot find the money on demand—has sometimes made the mortgage system approximate to a contract of land for future delivery against land for spot delivery. There have been sales of lands to tenants against mortgages effected by them, which, in fact, came very near to being transactions of this character.

That the world after several millennia of steady individual saving, is so poor as it is in accumulated capital-assets, is to be explained, in my opinion, neither by the improvident propensities of mankind, nor even by the destruction of war, but by the high liquidity-premiums formerly attaching to the ownership of land and now attaching to money. I differ in this from the older view as expressed by Marshall with an unusual dogmatic force in his *Principles of Economics*, p. 581:

> Everyone is aware that the accumulation of wealth is held in check, and the rate of interest so far sustained, by the preference which the great mass of humanity have for present over deferred gratifications, or, in other words, by their unwillingness to 'wait'.

VI

In my *Treatise on Money* I defined what purported to be a unique rate of interest, which I called the *natural rate* of interest—namely, the rate of interest which, in the terminology of my *Treatise*, preserved equality between the rate of saving (as there defined) and the rate of investment. I believed this to be a development and clarification of Wicksell's 'natural rate of interest', which was, according to him, the rate which would preserve the stability of some, not quite clearly specified, price-level.

I had, however, overlooked the fact that in any given society there is, on this definition, a *different* natural rate of interest for each hypothetical level of employment. And, similarly, for every rate of interest there is a level of employment for which that rate is the 'natural' rate, in the sense that the system will be in equilibrium with that rate of interest and that level of employment. Thus it was a mistake to speak of *the* natural rate of interest or to suggest that the above definition would yield a unique value for the rate of interest irrespective of the level of employment. I had not then understood that, in certain conditions, the system could be in equilibrium with less than full employment.

I am now no longer of the opinion that the concept of a 'natural' rate of interest, which previously seemed to me a most promising idea, has anything very useful or significant to contribute to our analysis. It is merely the rate of interest which will preserve the *status quo*; and, in general, we have no predominant interest in the *status quo* as such.

If there is any such rate of interest, which is unique and significant, it must be the rate which we might term the *neutral* rate of interest,[1] namely, the natural rate in the above sense which is consistent with *full* employment, given the other parameters of the system; though this rate might be better described, perhaps, as the *optimum* rate.

The neutral rate of interest can be more strictly defined as the rate of interest which prevails in equilibrium when output and employment are such that the elasticity of employment as a whole is zero.[2]

The above gives us, once again, the answer to the question as to what tacit assumption is required to make sense of the classical theory of the rate of interest. This theory assumes either that the actual rate of interest is always equal to the neutral rate of interest in the sense in which we have just defined the latter, or alternatively that the actual rate of interest is always equal to the rate of interest which will maintain employment at some specified constant level. If the traditional theory is thus interpreted, there is little or nothing in its practical conclusions to which we need take exception. The classical theory assumes that the banking authority or natural forces cause the market-rate of interest to

---

[1]   This definition does not correspond to any of the various definitions of *neutral money* given by recent writers; though it may, perhaps, have some relation to the objective which these writers have had in mind.
[2]   Cf. chapter 20 below.

satisfy one or other of the above conditions; and it investigates what laws will govern the application and rewards of the community's productive resources subject to this assumption. With this limitation in force, the volume of output depends solely on the assumed constant level of employment in conjunction with the current equipment and technique; and we are safely ensconced in a Ricardian world.

Chapter 18

THE GENERAL THEORY OF EMPLOYMENT RE-STATED

I

We have now reached a point where we can gather together the threads of our argument. To begin with, it may be useful to make clear which elements in the economic system we usually take as given, which are the independent variables of our system and which are the dependent variables.

We take as given the existing skill and quantity of available labour, the existing quality and quantity of available equipment, the existing technique, the degree of competition, the tastes and habits of the consumer, the disutility of different intensities of labour and of the activities of supervision and organisation, as well as the social structure including the forces, other than our variables set forth below, which determine the distribution of the national income. This does not mean that we assume these factors to be constant; but merely that, in this place and context, we are not considering or taking into account the effects and consequences of changes in them.

Our independent variables are, in the first instance, the propensity to consume, the schedule of the marginal efficiency of capital and the rate of interest, though, as we have already seen, these are capable of further analysis.

Our dependent variables are the volume of employment and the national income (or national dividend) measured in wage-units.

The factors, which we have taken as given, influence

our independent variables, but do not completely determine them. For example, the schedule of the marginal
efficiency of capital depends partly on the existing
quantity of equipment which is one of the given factors,
but partly on the state of long-term expectation which
cannot be inferred from the given factors. But there
are certain other elements which the given factors determine so completely that we can treat these derivatives
as being themselves given. For example, the given
factors allow us to infer what level of national income
measured in terms of the wage-unit will correspond to
any given level of employment; so that, within the
economic framework which we take as given, the
national income depends on the volume of employment,
i.e. on the quantity of effort currently devoted to production, in the sense that there is a unique correlation
between the two.[1] Furthermore, they allow us to infer
the shape of the aggregate supply functions, which
embody the *physical* conditions of supply, for different
types of products;—that is to say, the quantity of
employment which will be devoted to production corresponding to any given level of effective demand
measured in terms of wage-units. Finally, they furnish
us with the supply function of labour (or effort); so that
they tell us *inter alia* at what point the employment
function[2] for labour as a whole will cease to be elastic.

The schedule of the marginal efficiency of capital
depends, however, partly on the given factors and
partly on the prospective yield of capital-assets of
different kinds; whilst the rate of interest depends
partly on the state of liquidity-preference (i.e. on the
liquidity function) and partly on the quantity of money
measured in terms of wage-units. Thus we can sometimes regard our ultimate independent variables as con-
sisting of (1) the three fundamental psychological

[1] We are ignoring at this stage certain complications which arise when
the employment functions of different products have different curvatures
within the relevant range of employment. See chapter 20 below.
[2] Defined in chapter 20 below.

factors, namely, the psychological propensity to consume, the psychological attitude to liquidity and the psychological expectation of future yield from capital-assets, (2) the wage-unit as determined by the bargains reached between employers and employed, and (3) the quantity of money as determined by the action of the central bank; so that, if we take as given the factors specified above, these variables determine the national income (or dividend) and the quantity of employment. But these again would be capable of being subjected to further analysis, and are not, so to speak, our ultimate atomic independent elements.

The division of the determinants of the economic system into the two groups of given factors and independent variables is, of course, quite arbitrary from any absolute standpoint. The division must be made entirely on the basis of experience, so as to correspond on the one hand to the factors in which the changes seem to be so slow or so little relevant as to have only a small and comparatively negligible short-term influence on our *quaesitum*; and on the other hand to those factors in which the changes are found in practice to exercise a dominant influence on our *quaesitum*. Our present object is to discover what determines at any time the national income of a given economic system and (which is almost the same thing) the amount of its employment; which means in a study so complex as economics, in which we cannot hope to make completely accurate generalisations, the factors whose changes *mainly* determine our *quaesitum*. Our final task might be to select those variables which can be deliberately controlled or managed by central authority in the kind of system in which we actually live.

II

Let us now attempt to summarise the argument of the previous chapters; taking the factors in the reverse order to that in which we have introduced them.

There will be an inducement to push the rate of new investment to the point which forces the supply-price of each type of capital-asset to a figure which, taken in conjunction with its prospective yield, brings the marginal efficiency of capital in general to approximate equality with the rate of interest. That is to say, the physical conditions of supply in the capital-goods industries, the state of confidence concerning the prospective yield, the psychological attitude to liquidity and the quantity of money (preferably calculated in terms of wage-units) determine, between them, the rate of new investment.

But an increase (or decrease) in the rate of investment will have to carry with it an increase (or decrease) in the rate of consumption; because the behaviour of the public is, in general, of such a character that they are only willing to widen (or narrow) the gap between their income and their consumption if their income is being increased (or diminished). That is to say, changes in the rate of consumption are, in general, *in the same direction* (though smaller in amount) as changes in the rate of income. The relation between the increment of consumption which has to accompany a given increment of saving is given by the marginal propensity to consume. The ratio, thus determined, between an increment of investment and the corresponding increment of aggregate income, both measured in wage-units, is given by the investment multiplier.

Finally, if we assume (as a first approximation) that the employment multiplier is equal to the investment multiplier, we can, by applying the multiplier to the increment (or decrement) in the rate of investment brought about by the factors first described, infer the increment of employment.

An increment (or decrement) of employment is liable, however, to raise (or lower) the schedule of liquidity-preference; there being three ways in which it will tend to increase the demand for money, inasmuch as the value of output will rise when employment increases even if the wage-unit and prices (in terms of the wage-unit) are unchanged, but, in addition, the wage-unit itself will tend to rise as employment improves, and the increase in output will be accompanied by a rise of prices (in terms of the wage-unit) owing to increasing cost in the short period.

Thus the position of equilibrium will be influenced by these repercussions; and there are other repercussions also. Moreover, there is not one of the above factors which is not liable to change without much warning, and sometimes substantially. Hence the extreme complexity of the actual course of events. Nevertheless, these seem to be the factors which it is useful and convenient to isolate. If we examine any actual problem along the lines of the above schematism, we shall find it more manageable; and our practical intuition (which can take account of a more detailed complex of facts than can be treated on general principles) will be offered a less intractable material upon which to work.

III

The above is a summary of the General Theory. But the actual phenomena of the economic system are also coloured by certain special characteristics of the propensity to consume, the schedule of the marginal efficiency of capital and the rate of interest, about which we can safely generalise from experience, but which are not logically necessary.

In particular, it is an outstanding characteristic of the economic system in which we live that, whilst it is subject to severe fluctuations in respect of output and employment, it is not violently unstable. Indeed it seems capable of remaining in a chronic condition of sub-normal activity for a considerable period without any marked tendency either towards recovery or towards complete collapse. Moreover, the evidence indicates that full, or even approximately full, employment is of rare and short-lived occurrence. Fluctuations may start briskly but seem to wear themselves out before they have proceeded to great extremes, and an intermediate situation which is neither desperate nor satisfactory is our normal lot. It is upon the fact that fluctuations tend to wear themselves out before proceeding to extremes and eventually to reverse themselves, that the theory of business *cycles* having a regular phase has been founded. The same thing is true of prices, which, in response to an initiating cause of disturbance, seem to be able to find a level at which they can remain, for the time being, moderately stable.

Now, since these facts of experience do not follow of logical necessity, one must suppose that the environment and the psychological propensities of the modern world must be of such a character as to produce these results. It is, therefore, useful to consider what hypothetical psychological propensities would lead to a stable system; and, then, whether these propensities can be plausibly ascribed, on our general knowledge of contemporary human nature, to the world in which we live.

The conditions of stability which the foregoing analysis suggests to us as capable of explaining the observed results are the following:

(i) The marginal propensity to consume is such that, when the output of a given community increases (or decreases) because more (or less) employment is being applied to its capital equipment, the multiplier relating the two is greater than unity but not very large.

(ii) When there is a change in the prospective yield of capital or in the rate of interest, the schedule of the marginal efficiency of capital will be such that the change in new investment will not be in great disproportion to the change in the former; i.e. moderate changes in the prospective yield of capital or in the rate of interest will not be associated with very great changes in the rate of investment.

(iii) When there is a change in employment, money-wages tend to change in the same direction as, but not in great disproportion to, the change in employment; i.e. moderate changes in employment are not associated with very great changes in money-wages. This is a condition of the stability of prices rather than of employment.

(iv) We may add a fourth condition, which provides not so much for the stability of the system as for the tendency of a fluctuation in one direction to reverse itself in due course; namely, that a rate of investment, higher (or lower) than prevailed formerly, begins to react unfavourably (or favourably) on the marginal efficiency of capital if it is continued for a period which, measured in years, is not very large.

(i) Our first condition of stability, namely, that the multiplier, whilst greater than unity, is not very great, is highly plausible as a psychological characteristic of human nature. As real income increases, both the pressure of present needs diminishes and the margin over the established standard of life is increased; and as real income diminishes the opposite is true. Thus it is natural—at any rate on the average of the community —that current consumption should be expanded when employment increases, but by less than the full increment of real income; and that it should be diminished when employment diminishes, but by less than the full decrement of real income. Moreover, what is true of the average of individuals is likely to be also true of governments, especially in an age when a progressive increase of unemployment will usually force the State to provide relief out of borrowed funds.

But whether or not this psychological law strikes the reader as plausible *a priori*, it is certain that experience would be extremely different from what it is if the law did not hold. For in that case an increase of investment, however small, would set moving a cumulative increase of effective demand until a position of full employment had been reached; while a decrease of investment would set moving a cumulative decrease of effective demand until no one at all was employed. Yet experience shows that we are generally in an intermediate position. It is not impossible that there may be a range within which instability does in fact prevail. But, if so, it is probably a narrow one, outside of which in either direction our psychological law must unquestionably hold good. Furthermore, it is also evident that the multiplier, though exceeding unity, is not, in normal circumstances, enormously large. For, if it were, a given change in the rate of investment would involve a great change (limited only by full or zero employment) in the rate of consumption.

(ii) Whilst our first condition provides that a moderate change in the rate of investment will not involve an indefinitely great change in the demand for consumption-goods our second condition provides that a moderate change in the prospective yield of capital-assets or in the rate of interest will not involve an indefinitely great change in the rate of investment. This is likely to be the case owing to the increasing cost of producing a greatly enlarged output from the existing equipment. If, indeed, we start from a position where there are very large surplus resources for the production of capital-assets, there may be considerable instability within a certain range; but this will cease to hold good as soon as the surplus is being largely utilised. Moreover, this condition sets a limit to the instability resulting from rapid changes in the prospective yield of capital-assets due to sharp fluctuations in business psychology or to epoch-making inventions—though more, perhaps, in the upward than in the downward direction.

(iii) Our third condition accords with our experience of human nature. For although the struggle for money-wages is, as we have pointed out above, essentially a struggle to maintain a high *relative* wage, this struggle is likely, as employment increases, to be intensified in each individual case both because the bargaining position of the worker is improved and because the diminished marginal utility of his wage and his improved financial margin make him readier to run risks. Yet, all the same, these motives will operate within limits, and workers will not seek a much greater money-wage when employment improves or allow a very great reduction rather than suffer any unemployment at all.

But here again, whether or not this conclusion is plausible *a priori*, experience shows that some such psychological law must actually hold. For if competition between unemployed workers always led to a very great reduction of the money-wage, there would be a violent instability in the price-level. Moreover, there might be no position of stable equilibrium except in conditions consistent with full employment; since the wage-unit might have to fall without limit until it reached a point where the effect of the abundance of money in terms of the wage-unit on the rate of interest was sufficient to restore a level of full employment. At no other point could there be a resting-place.[1]

(iv) Our fourth condition, which is a condition not so much of stability as of alternate recession and recovery, is merely based on the presumption that capital-assets are of various ages, wear out with time and are not all very long-lived; so that if the rate of investment falls below a certain minimum level, it is merely a question of time (failing large fluctuations in other factors) before the marginal efficiency of capital rises sufficiently to bring about a recovery of investment above this minimum. And similarly, of course, if investment rises to a higher figure than formerly, it is only a question of time before the marginal efficiency of capital falls sufficiently to bring about

---

[1] The effects of changes in the wage-unit will be considered in detail in chapter 19.

a recession unless there are compensating changes in other factors.

For this reason, even those degrees of recovery and recession, which can occur within the limitations set by our other conditions of stability, will be likely, if they persist for a sufficient length of time and are not interfered with by changes in the other factors, to cause a reverse movement in the opposite direction, until the same forces as before again reverse the direction.

Thus our four conditions together are adequate to explain the outstanding features of our actual experience;—namely, that we oscillate, avoiding the gravest extremes of fluctuation in employment and in prices in both directions, round an intermediate position appreciably below full employment and appreciably above the minimum employment a decline below which would endanger life.

But we must not conclude that the mean position thus determined by 'natural' tendencies, namely, by those tendencies which are likely to persist, failing measures expressly designed to correct them, is, therefore, established by laws of necessity. The unimpeded rule of the above conditions is a fact of observation concerning the world as it is or has been, and not a necessary principle which cannot be changed.

BOOK V
MONEY-WAGES AND PRICES

Chapter 19

CHANGES IN MONEY-WAGES

I

It would have been an advantage if the effects of a change in money-wages could have been discussed in an earlier chapter. For the classical theory has been accustomed to rest the supposedly self-adjusting character of the economic system on an assumed fluidity of money-wages; and, when there is rigidity, to lay on this rigidity the blame of maladjustment.

It was not possible, however, to discuss this matter fully until our own theory had been developed. For the consequences of a change in money-wages are complicated. A reduction in money-wages is quite capable in certain circumstances of affording a stimulus to output, as the classical theory supposes. My difference from this theory is primarily a difference of analysis; so that it could not be set forth clearly until the reader was acquainted with my own method.

The generally accepted explanation is, as I understand it, quite a simple one. It does not depend on roundabout repercussions, such as we shall discuss below. The argument simply is that a reduction in money-wages will *cet. par.* stimulate demand by diminishing the price of the finished product, and will therefore increase output and employment up to the point where the reduction which labour has agreed to accept in its money-wages is just offset by the diminishing marginal efficiency of labour as output (from a given equipment) is increased.

In its crudest form, this is tantamount to assuming that the reduction in money-wages will leave demand unaffected. There may be some economists who would maintain that there is no reason why demand should be affected, arguing that aggregate demand depends on the quantity of money multiplied by the income-velocity of money and that there is no obvious reason why a reduction in money-wages would reduce either the quantity of money or its income-velocity. Or they may even argue that profits will necessarily go up because wages have gone down. But it would, I think, be more usual to agree that the reduction in money-wages may have *some* effect on aggregate demand through its reducing the purchasing power of some of the workers, but that the real demand of other factors, whose money incomes have not been reduced, will be stimulated by the fall in prices, and that the aggregate demand of the workers themselves will be very likely increased as a result of the increased volume of employment, unless the elasticity of demand for labour in response to changes in money-wages is less than unity. Thus in the new equilibrium there will be more employment than there would have been otherwise except, perhaps, in some unusual limiting case which has no reality in practice.

It is from this type of analysis that I fundamentally differ; or rather from the analysis which seems to lie behind such observations as the above. For whilst the above fairly represents, I think, the way in which many economists talk and write, the underlying analysis has seldom been written down in detail.

It appears, however, that this way of thinking is probably reached as follows. In any given industry we have a demand schedule for the product relating the quantities which can be sold to the prices asked; we have a series of supply schedules relating the prices which will be asked for the sale of different quantities on various bases of cost; and these schedules between them lead up to a further schedule which, on the assumption that other costs are unchanged (except as a result of the change in output), gives us the demand schedule for labour in the industry relating the quantity of employment to different levels of wages, the shape of the curve at any point furnishing the elasticity of demand for labour. This conception is then transferred without substantial modification to industry as a whole; and it is supposed, by a parity of reasoning, that we have a demand schedule for labour in industry as a whole relating the quantity of employment to different levels of wages. It is held that it makes no material difference to this argument whether it is in terms of money-wages or of real wages. If we are thinking in terms of money-wages, we must, of course, correct for changes in the value of money; but this leaves the general tendency of the argument unchanged, since prices certainly do not change in exact proportion to changes in money-wages.

If this is the groundwork of the argument (and, if it is not, I do not know what the groundwork is), surely it is fallacious. For the demand schedules for particular industries can only be constructed on some fixed assumption as to the nature of the demand and supply schedules of other industries and as to the amount of the aggregate effective demand. It is invalid, therefore, to transfer the argument to industry as a whole unless we also transfer our assumption that the aggregate effective demand is fixed. Yet this assumption reduces the argument to an *ignoratio elenchi*. For, whilst no one would wish to deny the proposition that a reduction in money-wages *accompanied by the same aggregate effective demand as before* will be associated with an increase in employment, the precise question at issue is whether the reduction in money-wages will or will not be accompanied by the same aggregate effective demand as before measured in money, or, at any rate, by an aggregate effective demand which is not reduced in full proportion to the reduction in money-wages (i.e. which is somewhat greater measured in wage-units). But if the classical theory is not allowed to extend by analogy its conclusions in respect of a particular industry to industry as a whole, it is wholly unable to answer the question what effect on employment a reduction in money-wages will have. For it has no method of analysis wherewith to tackle the problem. Professor Pigou's *Theory of Unemployment* seems to me to get out of the classical theory all that can be got out of it; with the result that the book becomes a striking demonstration that this theory has nothing to offer, when it is applied to the problem of what determines the volume of actual employment as a whole.[1]

II

Let us, then, apply our own method of analysis to answering the problem. It falls into two parts. (1) Does a reduction in money-wages have a direct tendency, *cet. par.*, to increase employment, '*cet. par.*' being taken to mean that the propensity to consume, the schedule of the marginal efficiency of capital and the rate of interest are the same as before for the community as a whole? And (2) does a reduction in money-wages have a certain or probable tendency to affect employment in a particular direction through its certain or probable repercussions on these three factors?

The first question we have already answered in the negative in the preceding chapters. For we have shown that the volume of employment is uniquely correlated with the volume of effective demand measured in wage-units, and that the effective demand, being the sum of the expected consumption and the expected investment, cannot change, if the propensity to consume, the schedule of marginal efficiency of capital and

---

[1] In an Appendix to this chapter Professor Pigou's *Theory of Unemployment* is criticised in detail.

the rate of interest are all unchanged. If, without any change in these factors, the entrepreneurs were to increase employment as a whole, their proceeds will necessarily fall short of their supply-price.

Perhaps it will help to rebut the crude conclusion that a reduction in money-wages will increase employment 'because it reduces the cost of production', if we follow up the course of events on the hypothesis most favourable to this view, namely that at the outset entrepreneurs *expect* the reduction in money-wages to have this effect. It is indeed not unlikely that the individual entrepreneur, seeing his own costs reduced, will overlook at the outset the repercussions on the demand for his product and will act on the assumption that he will be able to sell at a profit a larger output than before. If, then, entrepreneurs generally act on this expectation, will they in fact succeed in increasing their profits? Only if the community's marginal propensity to consume is equal to unity, so that there is no gap between the increment of income and the increment of consumption; or if there is an increase in investment, corresponding to the gap between the increment of income and the increment of consumption, which will only occur if the schedule of marginal efficiencies of capital has increased relatively to the rate of interest. Thus the proceeds realised from the increased output will disappoint the entrepreneurs and employment will fall back again to its previous figure, unless the marginal propensity to consume is equal to unity or the reduction in money-wages has had the effect of increasing the schedule of marginal efficiencies of capital relatively to the rate of interest and hence the amount of invest-ment. For if entrepreneurs offer employment on a scale which, if they could sell their output at the ex-pected price, would provide the public with incomes out of which they would save more than the amount of current investment, entrepreneurs are bound to make a loss equal to the difference; and this will be the case absolutely irrespective of the level of money-wages. At the best, the date of their disappointment can only be delayed for the interval during which their own investment in increased working capital is filling the gap.

Thus the reduction in money-wages will have no lasting tendency to increase employment except by virtue of its repercussion either on the propensity to consume for the community as a whole, or on the schedule of marginal efficiencies of capital, or on the rate of interest. There is no method of analysing the effect of a reduction in money-wages, except by following up its possible effects on these three factors.

The most important repercussions on these factors are likely, in practice, to be the following:

(1) A reduction of money-wages will somewhat reduce prices. It will, therefore, involve some redistribution of real income (*a*) from wage-earners to other factors entering into marginal prime cost whose remuneration has not been reduced, and (*b*) from entrepreneurs to rentiers to whom a certain income fixed in terms of money has been guaranteed.

What will be the effect of this redistribution on the propensity to consume for the community as a whole? The transfer from wage-earners to other factors is likely to diminish the propensity to consume. The effect of the transfer from entrepreneurs to rentiers is more open to doubt. But if rentiers represent on the whole the richer section of the community and those whose standard of life is least flexible, then the effect of this also will be unfavourable. What the net result will be on a balance of considerations, we can only guess. Probably it is more likely to be adverse than favourable.

(2) If we are dealing with an unclosed system, and the reduction of money-wages is a *reduction relatively to money-wages abroad* when both are reduced to a common unit, it is evident that the change will be favourable to investment, since it will tend to increase the balance of trade. This assumes, of course, that the

advantage is not offset by a change in tariffs, quotas, etc. The greater strength of the traditional belief in the efficacy of a reduction in money-wages as a means of increasing employment in Great Britain, as compared with the United States, is probably attributable to the latter being, comparatively with ourselves, a closed system.

(3) In the case of an unclosed system, a reduction of money-wages, though it increases the favourable balance of trade, is likely to worsen the terms of trade. Thus there will be a reduction in real incomes, except in the case of the newly employed, which may tend to increase the propensity to consume.

(4) If the reduction of money-wages is expected to be a *reduction relatively to money-wages in the future*, the change will be favourable to investment, because as we have seen above, it will increase the marginal efficiency of capital; whilst for the same reason it may be favourable to consumption. If, on the other hand, the reduction leads to the expectation, or even to the serious possibility, of a further wage-reduction in prospect, it will have precisely the opposite effect. For it will diminish the marginal efficiency of capital and will lead to the postponement both of investment and of consumption.

(5) The reduction in the wages-bill, accompanied by some reduction in prices and in money-incomes generally, will diminish the need for cash for income and business purposes; and it will therefore reduce *pro tanto* the schedule of liquidity-preference for the community as a whole. *Cet. par.* this will reduce the rate of interest and thus prove favourable to investment. In this case, however, the effect of expectation concerning the future will be of an opposite tendency to those just considered under (4). For, if wages and prices are expected to rise again later on, the favourable reaction will be much less pronounced in the case of long-term loans than in that of short-term loans. If, more-

over, the reduction in wages disturbs political confidence by causing popular discontent, the increase in liquidity-preference due to this cause may more than offset the release of cash from the active circulation.

(6) Since a special reduction of money-wages is always advantageous to an individual entrepreneur or industry, a general reduction (though its actual effects are different) may also produce an optimistic tone in the minds of entrepreneurs, which may break through a vicious circle of unduly pessimistic estimates of the marginal efficiency of capital and set things moving again on a more normal basis of expectation. On the other hand, if the workers make the same mistake as their employers about the effects of a general reduction, labour troubles may offset this favourable factor; apart from which, since there is, as a rule, no means of securing a simultaneous and equal reduction of money-wages in all industries, it is in the interest of all workers to resist a reduction in their own particular case. In fact, a movement by employers to revise money-wage bargains downward will be much more strongly resisted than a gradual and automatic lowering of real wages as a result of rising prices.

(7) On the other hand, the depressing influence on entrepreneurs of their greater burden of debt may partly offset any cheerful reactions from the reduction of wages. Indeed if the fall of wages and prices goes far, the embarrassment of those entrepreneurs who are heavily indebted may soon reach the point of insolvency, —with severely adverse effects on investment. Moreover the effect of the lower price-level on the real burden of the national debt and hence on taxation is likely to prove very adverse to business confidence.

This is not a complete catalogue of all the possible reactions of wage reductions in the complex real world. But the above cover, I think, those which are usually the most important.

If, therefore, we restrict our argument to the case

of a closed system, and assume that there is nothing to be hoped, but if anything the contrary, from the repercussions of the new distribution of real incomes on the community's propensity to spend, it follows that we must base any hopes of favourable results to employment from a reduction in money-wages mainly on an improvement in investment due either to an increased marginal efficiency of capital under (4) or a decreased rate of interest under (5). Let us consider these two possibilities in further detail.

The contingency, which is favourable to an increase in the marginal efficiency of capital, is that in which money-wages are believed to have touched bottom, so that further changes are expected to be in the upward direction. The most unfavourable contingency is that in which money-wages are slowly sagging downwards and each reduction in wages serves to diminish confidence in the prospective maintenance of wages. When we enter on a period of weakening effective demand, a sudden large reduction of money-wages to a level so low that no one believes in its indefinite continuance would be the event most favourable to a strengthening of effective demand. But this could only be accomplished by administrative decree and is scarcely practical politics under a system of free wage-bargaining. On the other hand, it would be much better that wages should be rigidly fixed and deemed incapable of material changes, than that depressions should be accompanied by a gradual downward tendency of money-wages, a further moderate wage reduction being expected to signalise each increase of, say, 1 per cent in the amount of unemployment. For example, the effect of an expectation that wages are going to sag by, say, 2 per cent in the coming year will be roughly equivalent to the effect of a rise of 2 per cent in the amount of interest payable for the same period. The same observations apply *mutatis mutandis* to the case of a boom.

It follows that with the actual practices and in-

stitutions of the contemporary world it is more expedient to aim at a rigid money-wage policy than at a
flexible policy responding by easy stages to changes in
the amount of unemployment;—so far, that is to say,
as the marginal efficiency of capital is concerned. But
is this conclusion upset when we turn to the rate of
interest?

It is, therefore, on the effect of a falling wage- and
price-level on the demand for money that those who
believe in the self-adjusting quality of the economic
system must rest the weight of their argument; though
I am not aware that they have done so. If the quantity
of money is itself a function of the wage- and price-
level, there is indeed, nothing to hope in this direction.
But if the quantity of money is virtually fixed, it is
evident that its quantity in terms of wage-units can
be indefinitely increased by a sufficient reduction in
money-wages; and that its quantity in proportion to
incomes generally can be largely increased, the limit
to this increase depending on the proportion of wage-
cost to marginal prime cost and on the response of
other elements of marginal prime cost to the falling
wage-unit.

We can, therefore, theoretically at least, produce
precisely the same effects on the rate of interest by
reducing wages, whilst leaving the quantity of money
unchanged, that we can produce by increasing the
quantity of money whilst leaving the level of wages
unchanged. It follows that wage reductions, as a
method of securing full employment, are also subject
to the same limitations as the method of increasing the
quantity of money. The same reasons as those mentioned above, which limit the efficacy of increases in
the quantity of money as a means of increasing investment to the optimum figure, apply *mutatis mutandis*
to wage reductions. Just as a moderate increase in
the quantity of money may exert an inadequate influence over the long-term rate of interest, whilst an

immoderate increase may offset its other advantages by its disturbing effect on confidence; so a moderate reduction in money-wages may prove inadequate, whilst an immoderate reduction might shatter confidence even if it were practicable.

There is, therefore, no ground for the belief that a flexible wage policy is capable of maintaining a state of continuous full employment;—any more than for the belief that an open-market monetary policy is capable, unaided, of achieving this result. The economic system cannot be made self-adjusting along these lines.

If, indeed, labour were always in a position to take action (and were to do so), whenever there was less than full employment, to reduce its money demands by concerted action to whatever point was required to make money so abundant relatively to the wage-unit that the rate of interest would fall to a level compatible with full employment, we should, in effect, have monetary management by the trade unions, aimed at full employment, instead of by the banking system.

Nevertheless while a flexible wage policy and a flexible money policy come, analytically, to the same thing, inasmuch as they are alternative means of changing the quantity of money in terms of wage-units, in other respects there is, of course, a world of difference between them. Let me briefly recall to the reader's mind the four outstanding considerations.

(i) Except in a socialised community where wage-policy is settled by decree, there is no means of securing uniform wage reductions for every class of labour. The result can only be brought about by a series of gradual, irregular changes, justifiable on no criterion of social justice or economic expedience, and probably completed only after wasteful and disastrous struggles, where those in the weakest bargaining position will suffer relatively to the rest. A change in the quantity of money, on the other hand, is already within the power of most governments by open-market policy or analogous measures. Having regard to human nature and our institutions, it can only be a foolish person who would prefer a flexible wage policy to a flexible money policy, unless he can point to advantages from the former which are not obtainable from the latter. Moreover, other things being equal, a method which it is comparatively easy to apply should be deemed preferable to a method which is probably so difficult as to be impracticable.

(ii) If money-wages are inflexible, such changes in prices as occur (i.e. apart from 'administered' or monopoly prices which are determined by other considerations besides marginal cost) will mainly correspond to the diminishing marginal productivity of the existing equipment as the output from it is increased. Thus the greatest practicable fairness will be maintained between labour and the factors whose remuneration is contractually fixed in terms of money, in particular the rentier class and persons with fixed salaries on the permanent establishment of a firm, an institution or the State. If important classes are to have their remuneration fixed in terms of money in any case, social justice and social expediency are best served if the remunerations of *all* factors are somewhat inflexible in terms of money. Having regard to the large groups of incomes which are comparatively inflexible in terms of money, it can only be an unjust person who would prefer a flexible wage policy to a flexible money policy, unless he can point to advantages from the former which are not obtainable from the latter.

(iii) The method of increasing the quantity of money in terms of wage-units by decreasing the wage-unit increases proportionately the burden of debt; whereas the method of producing the same result by increasing the quantity of money whilst leaving the wage-unit unchanged has the opposite effect. Having regard to the excessive burden of many types of debt, it can only be an inexperienced person who would prefer the former.

(iv) If a sagging rate of interest has to be brought about by a sagging wage-level, there is, for the reasons given above, a double drag on the marginal efficiency of capital and a double reason for putting off investment and thus postponing recovery.

III

It follows, therefore, that if labour were to respond to conditions of gradually diminishing employment by offering its services at a gradually diminishing money-wage, this would not, as a rule, have the effect of reducing real wages and might even have the effect of increasing them, through its adverse influence on the volume of output. The chief result of this policy would be to cause a great instability of prices, so violent perhaps as to make business calculations futile in an economic society functioning after the manner of that in which we live. To suppose that a flexible wage policy is a right and proper adjunct of a system which on the whole is one of *laissez-faire*, is the opposite of the truth. It is only in a highly authoritarian society, where sudden, substantial, all-round changes could be decreed that a flexible wage policy could function with success. One can imagine it in operation in Italy, Germany or Russia, but not in France, the United States or Great Britain.

If, as in Australia, an attempt were made to fix real wages by legislation, then there would be a certain level of employment corresponding to that level of real wages; and the actual level of employment would, in a closed system, oscillate violently between that level and no employment at all, according as the rate of investment was or was not below the rate compatible with that level; whilst prices would be in unstable equilibrium when investment was at the critical level, racing to zero whenever investment was below it, and to infinity whenever it was above it. The element of stability would have to be found, if at all, in the factors controlling the quantity of money being so determined that there always existed some level of money-wages at which the quantity of money would be such as to establish a relation between the rate of interest and the marginal efficiency of capital which would maintain investment at the critical level. In this event employment would be constant (at the level appropriate to the legal real wage) with money-wages and prices fluctuating rapidly in the degree just necessary to maintain this rate of investment at the appropriate figure. In the actual case of Australia, the escape was found, partly of course in the inevitable inefficacy of the legislation to achieve its object, and partly in Australia not being a closed system, so that the level of money-wages was itself a determinant of the level of foreign investment and hence of total investment, whilst the terms of trade were an important influence on real wages.

In the light of these considerations I am now of the opinion that the maintenance of a stable general level of money-wages is, on a balance of considerations, the most advisable policy for a closed system; whilst the same conclusion will hold good for an open system, provided that equilibrium with the rest of the world can be secured by means of fluctuating exchanges. There are advantages in some degree of flexibility in the wages of particular industries so as to expedite transfers from those which are relatively declining to those which are relatively expanding. But the money-wage level as a whole should be maintained as stable as possible, at any rate in the short period.

This policy will result in a fair degree of stability in the price-level;—greater stability, at least, than with a flexible wage policy. Apart from 'administered' or monopoly prices, the price-level will only change in the short period in response to the extent that changes in the volume of employment affect marginal prime costs; whilst in the long period they will only change in response to changes in the cost of production due to new techniques and new or increased equipment.

It is true that, if there are, nevertheless, large fluctuations in employment, substantial fluctuations in the price-level will accompany them. But the fluctuations will be less, as I have said above, than with a flexible wage policy.

Thus with a rigid wage policy the stability of prices will be bound up in the short period with the avoidance of fluctuations in employment. In the long period, on the other hand, we are still left with the choice between a policy of allowing prices to fall slowly with the progress of technique and equipment whilst keeping wages stable, or of allowing wages to rise slowly whilst keeping prices stable. On the whole my preference is for the latter alternative, on account of the fact that it is easier with an expectation of higher wages in future to keep the actual level of employment within a given range of full employment than with an expectation of lower wages in future, and on account also of the social advantages of gradually diminishing the burden of debt, the greater ease of adjustment from decaying to growing industries, and the psychological encouragement likely to be felt from a moderate tendency for money-wages to increase. But no essential point of principle is involved, and it would lead me beyond the scope of my present purpose to develop in detail the arguments on either side.

Appendix to Chapter 19

PROFESSOR PIGOU'S 'THEORY OF UNEMPLOYMENT'

Professor Pigou in his *Theory of Unemployment* makes the volume of employment to depend on two fundamental factors, namely (1) the real rates of wages for which workpeople stipulate, and (2) the shape of the Real Demand Function for Labour. The central sections of his book are concerned with determining the shape of the latter function. The fact that workpeople in fact stipulate, not for a real rate of wages, but for a money-rate, is not ignored; but, in effect, it is assumed that the actual money-rate of wages divided by the price of wage-goods can be taken to measure the real rate demanded.

The equations which, as he says, 'form the starting point of the enquiry' into the Real Demand Function for Labour are given in his *Theory of Unemployment*, p. 90. Since the tacit assumptions, which govern the application of his analysis, slip in near the outset of his argument, I will summarise his treatment up to the crucial point.

Professor Pigou divides industries into those 'engaged in making wage-goods at home and in making exports the sale of which creates claims to wage-goods abroad' and the 'other' industries: which it is convenient to call the wage-goods industries and the non-wage-goods industries respectively. He supposes x men to be employed in the former and y men in the latter. The output in value of wage-goods of the x men he calls $F(x)$; and the general rate of wages $F'(x)$. This, though he does not stop to mention it, is tantamount to assuming that marginal wage-cost is equal to marginal prime cost.[1] Further, he assumes

---

[1] The source of the fallacious practice of equating marginal wage cost to marginal prime cost may, perhaps, be found in an ambiguity in the meaning of *marginal wage cost*. We might mean by it the cost of an additional unit of output if no additional cost is incurred except additional wage-cost; or we might mean the additional wage cost involved in producing an additional unit of output in the most economical way with the help of the existing equipment and other unemployed factors. In the former case we are pre-

272

that $x+y = \phi(x)$, i.e. that the number of men employed in the wage-goods industries is a function of total employment. He then shows that the elasticity of the real demand for labour in the aggregate (which gives us the shape of our *quaesitum*, namely the Real Demand Function for Labour) can be written

$$E_r = \frac{\phi'(x)}{\phi(x)} \cdot \frac{F'(x)}{F''(x)}.$$

So far as notation goes, there is no significant difference between this and my own modes of expression. In so far as we can identify Professor Pigou's wage-goods with my consumption-goods, and his 'other goods' with my investment-goods, it follows that his $\frac{F(x)}{F'(x)}$, being the value of the output of the wage-goods industries in terms of the wage-unit, is the same as my C_w. Furthermore, his function ϕ is (subject to the identification of wage-goods with consumption-goods) a function of what I have called above the employment multiplier k'. For

$$\Delta x = k'\Delta y,$$

so that $$\phi'(x) = 1 + \frac{1}{k'}.$$

Thus Professor Pigou's 'elasticity of the real demand for labour in the aggregate' is a concoction similar to some of my own, depending partly on the physical and technical conditions in industry (as given by his function F) and partly on the propensity to consume wage-goods (as given by his function ϕ); provided always that we are limiting ourselves to the special case where marginal labour-cost is equal to marginal prime cost.

To determine the quantity of employment, Professor Pigou

cluded from combining with the additional labour any additional entre-preneurship or working capital or anything else other than labour which would add to the cost; and we are even precluded from allowing the additional labour to wear out the equipment any faster than the smaller labour force would have done. Since in the former case we have forbidden any element of cost other than labour cost to enter into marginal prime-cost, it does, of course, follow that marginal wage-cost and marginal prime-cost are equal. But the results of an analysis conducted on this premiss have almost no application, since the assumption on which it is based is very seldom realised in practice. For we are not so foolish in practice as to refuse to associate with additional labour appropriate additions of other factors, in so far as they are available, and the assumption will, therefore, only apply if we assume that all the factors, other than labour, are already being employed to the utmost.

then combines with his 'real demand for labour', a supply function for labour. He assumes that this is a function of the real wage and of nothing else. But, as he has also assumed that the real wage is a function of the number of men x who are employed in the wage-goods industries, this amounts to assuming that the total supply of labour at the existing real wage is a function of x and of nothing else. That is to say, $n = \chi(x)$, where n is the supply of labour available at a real wage $F'(x)$.

Thus, cleared of all complication, Professor Pigou's analysis amounts to an attempt to discover the volume of actual employment from the equations

$$x+y = \phi(x)$$

and

$$n = \chi(x).$$

But there are here three unknowns and only two equations. It seems clear that he gets round this difficulty by taking $n = x+y$. This amounts, of course, to assuming that there is no involuntary unemployment in the strict sense, i.e. that all labour available at the existing real wage is in fact employed. In this case x has the value which satisfies the equation

$$\phi(x) = \chi(x);$$

and when we have thus found that the value of x is equal to (say) n_1, y must be equal to $\chi(n_1) - n_1$, and total employment n is equal to $\chi(n_1)$.

It is worth pausing for a moment to consider what this involves. It means that, if the supply function of labour changes, more labour being available at a given real wage (so that $n_1 + dn_1$ is now the value of x which satisfies the equation $\phi(x) = \chi(x)$), the demand for the output of the non-wage-goods industries is such that employment in these industries is bound to increase by just the amount which will preserve equality between $\phi(n_1 + dn_1)$ and $\chi(n_1 + dn_1)$. The only other way in which it is possible for aggregate employment to change is through a modification of the propensity to purchase wage-goods and non-wage-goods respectively such that there is an increase of y accompanied by a greater decrease of x.

The assumption that $n = x+y$ means, of course, that labour is always in a position to determine its own real wage. Thus, the assumption that labour is in a position to determine its own real wage, means that the demand for the output of the non-wage-goods industries obeys the above laws. In other words, it is assumed that the rate of interest always adjusts itself to the schedule of the marginal efficiency of capital in such a way as to preserve full employment. Without this assumption Professor Pigou's analysis breaks down and provides no means of determining what the volume of employment will be. It is, indeed, strange that Professor Pigou should have supposed that he could furnish a theory of unemployment which involves no reference at all to changes in the rate of investment (i.e. to changes in employment in the non-wage-goods industries) due, not to a change in the supply function of labour, but to changes in (e.g.) either the rate of interest or the state of confidence.

His title the 'Theory of Unemployment' is, therefore, something of a misnomer. His book is not really concerned with this subject. It is a discussion of how much employment there will be, given the supply function of labour, when the conditions for full employment are satisfied. The purpose of the concept of the elasticity of the real demand for labour in the aggregate is to show by how much *full* employment will rise or fall corresponding to a given shift in the supply function of labour. Or—alternatively and perhaps better—we may regard his book as a non-causative investigation into the functional relationship which determines what level of real wages will correspond to any given level of employment. But it is not capable of telling us what determines the *actual* level of employment; and on the problem of involuntary unemployment it has no direct bearing.

If Professor Pigou were to deny the possibility of involuntary unemployment in the sense in which I have defined it above, as, perhaps, he would, it is still difficult to see how his analysis could be applied. For his omission to discuss what determines the connection between x and y, i.e. between employment in the wage-goods and non-wage-goods industries respectively, still remains fatal.

Moreover, he agrees that within certain limits labour in fact often stipulates, not for a given real wage, but for a given money-wage. But in this case the supply function of labour is not a function of $F'(x)$ alone but also of the money-price of wage-goods;—with the result that the previous analysis breaks down and an additional factor has to be introduced, without there being an additional equation to provide for this additional unknown. The pitfalls of a pseudo-mathematical method, which can make no progress except by making everything a function of a single variable and assuming that all the partial differentials vanish, could not be better illustrated. For it is no good to admit later on that there are in fact other variables, and yet to proceed without re-writing everything that has been written up to that point. Thus if (within limits) it is a money-wage for which labour stipulates, we still have insufficient data, even if we assume that $n = x + y$, unless we know what determines the money-price of wage-goods. For, the money-price of wage-goods will depend on the aggregate amount of employment. Therefore we cannot say what aggregate employment will be, until we know the money-price of wage-goods; and we cannot know the money-price of wage-goods until we know the aggregate amount of employment. We are, as I have said, one equation short. Yet it might be a provisional assumption of a rigidity of money-wages, rather than of real wages, which would bring our theory nearest to the facts. For example, money-wages in Great Britain during the turmoil and uncertainty and wide price fluctuations of the decade 1924–1934 were stable within a range of 6 per cent, whereas real wages fluctuated by more than 20 per cent. A theory cannot claim to be a *general* theory, unless it is applicable to the case where (or the range within which) money-wages are fixed, just as much as to any other case. Politicians are entitled to complain that money-wages *ought* to be highly flexible; but a theorist must be prepared to deal indifferently with either state of affairs. A scientific theory cannot require the facts to conform to its own assumptions.

When Professor Pigou comes to deal expressly with the effect of a reduction of money-wages, he again, palpably (to my mind), introduces too few data to permit of any definite answer being obtainable. He begins by rejecting the argument (*op. cit.* p. 101) that, if marginal prime cost is equal to marginal wage-cost, non-wage-earners' incomes will be altered, when money-wages are reduced, in the same proportion as wage-earners', on the ground that this is only valid, *if* the quantity of employment remains unaltered—which is the very point under discussion. But he proceeds on the next page (*op. cit.* p. 102) to make the same mistake himself by taking as his assumption that 'at the outset nothing has happened to non-wage-earners' money-income', which, as he has just shown, is only valid *if* the quantity of employment does *not* remain unaltered—which is the very point under discussion. In fact, *no* answer is possible, unless other factors are included in our data.

The manner in which the admission, that labour in fact stipulates for a given money-wage and not for a given real wage (provided that the real wage does not fall below a certain minimum), affects the analysis, can also be shown by pointing out that in this case the assumption that more labour is not available except at a greater real wage, which is fundamental to most of the argument, breaks down. For example, Professor Pigou rejects (*op. cit.* p. 75) the theory of the multiplier by assuming that the rate of real wages is given, i.e. that, there being already full employment, no additional labour is forthcoming at a lower real wage. Subject to this assumption, the argument is, of course, correct. But in this passage Professor Pigou is criticising a proposal relating to practical policy; and it is fantastically far removed from the facts to assume, at a time when statistical unemployment in Great Britain exceeded 2,000,000 (i.e. when there were 2,000,000 men willing to work at the existing money-wage), that any rise in the cost of living, however moderate, relatively to the money-wage would cause the withdrawal from the labour market of more than the equivalent of all these 2,000,000 men.

It is important to emphasise that the whole of Professor Pigou's book is written on the assumption *that any rise in the cost of living, however moderate, relatively to the money-wage will cause the withdrawal from the labour market of a number of workers greater than that of all the existing unemployed.*

Moreover, Professor Pigou does not notice in this passage (*op. cit.* p. 75) that the argument, which he advances against 'secondary' employment as a result of public works, is, on the same assumptions, equally fatal to increased 'primary' employment from the same policy. For if the real rate of wages ruling in the wage-goods industries is given, no increased employment whatever is possible—except, indeed, as a result of non-wage-earners reducing their consumption of wage-goods. For those newly engaged in the primary employment will presumably increase their consumption of wage-goods which will reduce the real wage and hence (on his assumptions) lead to a withdrawal of labour previously employed elsewhere. Yet Professor Pigou accepts, apparently, the possibility of increased primary employment. The line between primary and secondary employment seems to be the critical psychological point at which his good common sense ceases to overbear his bad theory.

The difference in the conclusions to which the above differences in assumptions and in analysis lead can be shown by the following important passage in which Professor Pigou sums up his point of view: 'With perfectly free competition among workpeople and labour perfectly mobile, the nature of the relation (i.e. between the real wage-rates for which people stipulate and the demand function for labour) will be very simple. There will always be at work a strong tendency for wage-rates to be so related to demand that everybody is employed. Hence, in stable conditions everyone will actually be employed. The implication is that such unemployment as exists at any time is due wholly to the fact that changes in demand conditions are continually taking place and that frictional resistances prevent the appropriate wage adjustments from being made instantaneously.'[1]

He concludes (*op. cit.* p. 253) that unemployment is primarily due to a wage policy which fails to adjust itself sufficiently to changes in the real demand function for labour.

Thus Professor Pigou believes that in the long run unemployment can be cured by wage adjustments;[2] whereas I maintain that the real wage (subject only to a minimum set by the marginal disutility of employment) is not primarily determined by 'wage adjustments' (though these may have repercussions) but by the other forces of the system, some of which (in particular the relation between the schedule of the marginal efficiency of capital and the rate of interest) Professor Pigou has failed, if I am right, to include in his formal scheme.

Finally, when Professor Pigou comes to the 'Causation of Unemployment' he speaks, it is true, of fluctuations in the state of demand, much as I do. But he identifies the state of demand with the Real Demand Function for Labour, forgetful of how narrow a thing the latter is on his definition. For the Real Demand Function for Labour depends by definition (as we have seen above) on *nothing* but two factors, namely (1) the relationship in any given environment between the total number of men employed and the number who have to be employed in the wage-goods industries to provide them with what they consume, and (2) the state of marginal productivity in the wage-goods industries. Yet in Part V of his *Theory of Unemployment* fluctuations in the state of 'the real demand for labour' are given a position of importance. The 'real demand for labour' is regarded as a factor which is susceptible of wide short-period fluctuations (*op. cit.* Part V, chaps. vi.-xii.), and the suggestion seems to be that swings in 'the real demand for labour' are, in combination with the failure of wage policy to respond sensitively to such changes, largely responsible for the trade cycle. To the reader all this seems, at first, reasonable and familiar. For, unless he goes back to the definition, 'fluctuations in the real demand for labour' will convey to his mind the same sort of suggestion as I mean to convey by 'fluctuations in the state of aggregate demand'. But if we go back to the definition of the 'real demand for labour', all this

[1] *Op. cit.* p. 252.
[2] There is no hint or suggestion that this comes about through reactions on the rate of interest.

loses its plausibility. For we shall find that there is nothing in the world less likely to be subject to sharp short-period swings than this factor.

Professor Pigou's 'real demand for labour' depends, by definition, on nothing but $F(x)$, which represents the physical conditions of production in the wage-goods industries, and $\phi(x)$, which represents the functional relationship between employment in the wage-goods industries and total employment corresponding to any given level of the latter. It is difficult to see a reason why either of these functions should change, except gradually over a long period. Certainly there seems no reason to suppose that they are likely to fluctuate during a trade cycle. For $F(x)$ can only change slowly, and, in a technically progressive community, only in the forward direction; whilst $\phi(x)$ will remain stable, unless we suppose a sudden outbreak of thrift in the working classes, or, more generally, a sudden shift in the propensity to consume. I should expect, therefore, that the real demand for labour would remain virtually constant throughout a trade cycle. I repeat that Professor Pigou has altogether omitted from his analysis the unstable factor, namely fluctuations in the scale of investment, which is most often at the bottom of the phenomenon of fluctuations in employment.

I have criticised at length Professor Pigou's theory of unemployment not because he seems to me to be more open to criticism than other economists of the classical school; but because his is the only attempt with which I am acquainted to write down the classical theory of unemployment precisely. Thus it has been incumbent on me to raise my objections to this theory in the most formidable presentment in which it has been advanced.

Chapter 20

THE EMPLOYMENT FUNCTION[1]

I

In chapter 3 (p. 23) we have defined the aggregate supply function $Z = \phi(N)$, which relates the employment N with the aggregate supply price of the corresponding output. The *employment function* only differs from the aggregate supply function in that it is, in effect, its inverse function and is defined in terms of the wage-unit; the object of the employment function being to relate the amount of the effective demand, measured in terms of the wage-unit, directed to a given firm or industry or to industry as a whole with the amount of employment, the supply price of the output of which will compare to that amount of effective demand. Thus if an amount of effective demand D_{wr}, measured in wage-units, directed to a firm or industry calls forth an amount of employment N_r in that firm or industry, the employment function is given by $N_r = F_r(D_{wr})$. Or, more generally, if we are entitled to assume that D_{wr} is a unique function of the total effective demand D_w, the employment function is given by $N_r = F_r(D_w)$. That is to say, N_r men will be employed in industry r when effective demand is D_w.

We shall develop in this chapter certain properties of the employment function. But apart from any interest which these may have, there are two reasons why the substitution of the employment function for

[1] Those who (rightly) dislike algebra will lose little by omitting the first section of this chapter.

the ordinary supply curve is consonant with the methods and objects of this book. In the first place, it expresses the relevant facts in terms of the units to which we have decided to restrict ourselves, without introducing any of the units which have a dubious quantitative character. In the second place, it lends itself to the problems of industry and output *as a whole*, as distinct from the problems of a single industry or firm in a given environment, more easily than does the ordinary supply curve—for the following reasons.

The ordinary demand curve for a particular commodity is drawn on some assumption as to the incomes of members of the public, and has to be re-drawn if the incomes change. In the same way the ordinary supply curve for a particular commodity is drawn on some assumption as to the output of industry as a whole and is liable to change if the aggregate output of industry is changed. When, therefore, we are examining the response of individual industries to changes in *aggregate* employment, we are necessarily concerned, not with a single demand curve for each industry, in conjunction with a single supply curve, but with two families of such curves corresponding to different assumptions as to the aggregate employment. In the case of the employment function, however, the task of arriving at a function for industry as a whole which will reflect changes in employment as a whole is more practicable.

For let us assume (to begin with) that the propensity to consume is given as well as the other factors which we have taken as given in chapter 18 above, and that we are considering changes in employment in response to changes in the rate of investment. Subject to this assumption, for every level of effective demand in terms of wage-units there will be a corresponding aggregate employment and this effective demand will be divided in determinate proportions between consumption and investment. Moreover, each level of effective demand will correspond to a given distribution of income. It is reasonable, therefore, further to assume that corresponding to a given level of aggregate effective demand there is a unique distribution of it between different industries.

This enables us to determine what amount of employment in each industry will correspond to a given level of aggregate employment. That is to say, it gives us the amount of employment in each particular industry corresponding to each level of aggregate effective demand measured in terms of wage-units, so that the conditions are satisfied for the second form of the employment function for the industry, defined above, namely $N_r = F_r(D_w)$. Thus we have the advantage that, in these conditions, the individual employment functions are additive in the sense that the employment function for industry as a whole, corresponding to a given level of effective demand, is equal to the sum of the employment functions for each separate industry; i.e.

$$F(D_w) = N = \Sigma N_r = \Sigma F_r(D_w).$$

Next, let us define the elasticity of employment. The elasticity of employment for a given industry is

$$e_{er} = \frac{dN_r}{dD_{wr}} \frac{D_{wr}}{N_r},$$

since it measures the response of the number of labour-units employed in the industry to changes in the number of wage-units which are expected to be spent on purchasing its output. The elasticity of employment for industry as a whole we shall write

$$e_e = \frac{dN}{dD_w} \cdot \frac{D_w}{N}.$$

Provided that we can find some sufficiently satisfactory method of measuring output, it is also useful to define what may be called the elasticity of output or production, which measures the rate at which output in any industry increases when more effective demand in terms of wage-units is directed towards it, namely

$$e_{or} = \frac{dO_r}{dD_{wr}} \cdot \frac{D_{wr}}{O_r}.$$

Provided we can assume that the price is equal to the marginal prime cost, we then have

$$\Delta D_{wr} = \frac{1}{1 - e_{or}} \Delta P_r$$

where P_r is the expected profit.[1] It follows from this that if $e_{or} = 0$, i.e. if the output of the industry is perfectly inelastic, the whole of the increased effective demand (in terms of wage-units) is expected to accrue to the entrepreneur as profit, i.e. $\Delta D_{wr} = \Delta P_r$; whilst if $e_{or} = 1$, i.e. if the elasticity of output is unity, no part of the increased effective demand is expected to accrue as profit, the whole of it being absorbed by the elements entering into marginal prime cost.

Moreover, if the output of an industry is a function $\phi(N_r)$ of the labour employed in it, we have[2]

$$\frac{1 - e_{or}}{e_{er}} = -\frac{N_r \phi''(N_r)}{p_{wr}\{\phi'(N_r)\}^2},$$

where p_{wr} is the expected price of a unit of output in

---

[1] For, if p_{wr} is the expected price of a unit of output in terms of the wage-unit,

$$\Delta D_{wr} = \Delta(p_{wr}O_r) = p_{wr}\Delta O_r + O_r\Delta p_{wr}$$
$$= \frac{D_{wr}}{O_r}.\Delta O_r + O_r\Delta p_{wr},$$

so that
$$O_r\Delta p_{wr} = \Delta D_{wr}(1 - e_{or})$$

or
$$\Delta D_{wr} = \frac{O_r\Delta p_{wr}}{1 - e_{or}}.$$

But
$$O_r\Delta p_{wr} = \Delta D_{wr} - p_{wr}\Delta O_r$$
$$= \Delta D_{wr} - (\text{marginal prime cost})\, \Delta O$$
$$= \Delta P.$$

Hence
$$\Delta D_{wr} = \frac{1}{1 - e_{or}}\Delta P_r.$$

[2] For, since $D_{wr} = p_{wr}O_r$, we have

$$1 = p_{wr}\frac{dO_r}{dD_{wr}} + O_r\frac{dp_{wr}}{dD_{wr}}$$
$$= e_{or} - \frac{N_r\phi''(N_r)\, e_{er}}{\{\phi'(N_r)\}^2\, p_{wr}}.$$

terms of the wage-unit. Thus the condition $e_{or} = 1$ means that $\phi''(N_r) = 0$, i.e. that there are constant returns in response to increased employment.

Now, in so far as the classical theory assumes that real wages are always equal to the marginal disutility of labour and that the latter increases when employment increases, so that the labour supply will fall off, *cet. par.*, if real wages are reduced, it is assuming that in practice it is impossible to increase expenditure in terms of wage-units. If this were true, the concept of elasticity of employment would have no field of application. Moreover, it would, in this event, be impossible to increase employment by increasing expenditure in terms of money; for money-wages would rise proportionately to the increased money expenditure so that there would be no increase of expenditure in terms of wage-units and consequently no increase in employment. But if the classical assumption does not hold good, it will be possible to increase employment by increasing expenditure in terms of money until real wages have fallen to equality with the marginal disutility of labour, at which point there will, by definition, be full employment.

Ordinarily, of course, e_{or} will have a value intermediate between zero and unity. The extent to which prices (in terms of wage-units) will rise, i.e. the extent to which real wages will fall, when money expenditure is increased, depends, therefore, on the elasticity of output in response to expenditure in terms of wage-units.

Let the elasticity of the expected price p_{wr} in response to changes in effective demand D_{wr}, namely $\dfrac{dp_{wr}}{dD_{wr}} \cdot \dfrac{D_{wr}}{p_{wr}}$, be written e'_{pr}.

Since $O_r \cdot p_{wr} = D_{wr}$, we have

$$\frac{dO_r}{dD_{wr}} \cdot \frac{D_{wr}}{O_r} + \frac{dp_{wr}}{dD_{wr}} \cdot \frac{D_{wr}}{p_{wr}} = 1$$

or

$$e'_{pr} + e_{or} = 1.$$

That is to say, the sum of the elasticities of price and of output in response to changes in effective demand (measured in terms of wage-units) is equal to unity. Effective demand spends itself, partly in affecting output and partly in affecting price, according to this law.

If we are dealing with industry as a whole and are prepared to assume that we have a unit in which output as a whole can be measured, the same line of argument applies, so that $e'_p + e_o = 1$, where the elasticities without a suffix r apply to industry as a whole.

Let us now measure values in money instead of wage-units and extend to this case our conclusions in respect of industry as a whole.

If W stands for the money-wages of a unit of labour and p for the expected price of a unit of output as a whole in terms of money, we can write $e_p \left(= \dfrac{D dp}{p dD} \right)$ for the elasticity of money-prices in response to changes in effective demand measured in terms of money, and $e_w \left(= \dfrac{D dW}{W dD} \right)$ for the elasticity of money-wages in response to changes in effective demand in terms of money. It is then easily shown that

$$e_p = 1 = e_o(1 - e_w).\text{[1]}$$

This equation is, as we shall see in the next chapter, a first step to a generalised quantity theory of money.

---

[1] For, since $p = p_w.W$ and $D = D_w.W$, we have

$$\Delta p = W \Delta p_w + \frac{p}{W} \Delta W$$

$$= W.e'_p \frac{p_w}{D_w} \Delta D_w + \frac{p}{W} \Delta W$$

$$= e'_p \frac{p}{D} \left(\Delta D - \frac{D}{W} \Delta W \right) + \frac{p}{W} \Delta W$$

$$= e'_p \frac{p}{D} \Delta D + \Delta W \frac{p}{W}(1 - e'_p),$$

so that
$$e_p = \frac{D \Delta p}{p \Delta D} = e'_p + \frac{D}{p \Delta D} \cdot \frac{\Delta W.p}{W}(1 - e'_p)$$

$$= e'_p + e_w(1 - e'_p)$$

$$= 1 - e_o(1 - e_w).$$

If $e_o = 0$ or if $e_w = 1$, output will be unaltered and prices will rise in the same proportion as effective demand in terms of money. Otherwise they will rise in a smaller proportion.

II

Let us return to the employment function. We have assumed in the foregoing that to every level or aggregate effective demand there corresponds a unique distribution of effective demand between the products of each individual industry. Now, as aggregate expenditure changes, the corresponding expenditure on the products of an individual industry will not, in general, change in the same proportion;—partly because individuals will not, as their incomes rise, increase the amount of the products of each separate industry, which they purchase, in the same proportion, and partly because the prices of different commodities will respond in different degrees to increases in expenditure upon them.

It follows from this that the assumption upon which we have worked hitherto, that changes in employment depend solely on changes in aggregate effective demand (in terms of wage-units), is no better than a first approximation, if we admit that there is more than one way in which an increase of income can be spent. For the way in which we suppose the increase in aggregate demand to be distributed between different commodities may considerably influence the volume of employment. If, for example, the increased demand is largely directed towards products which have a high elasticity of employment, the aggregate increase in employment will be greater than if it is largely directed towards products which have a low elasticity of employment.

In the same way employment may fall off without there having been any change in aggregate demand, if the direction of demand is changed in favour of products having a relatively low elasticity of employment.

These considerations are particularly important if we are concerned with short-period phenomena in the sense of changes in the amount or direction of demand which are not foreseen some time ahead. Some products take time to produce, so that it is practically impossible to increase the supply of them quickly. Thus, if additional demand is directed to them without notice, they will show a low elasticity of employment; although it may be that, given sufficient notice, their elasticity of employment approaches unity.

It is in this connection that I find the principal significance of the conception of a period of production. A product, I should prefer to say,[1] has a period of production n if n time-units of notice of changes in the demand for it have to be given if it is to offer its maximum elasticity of employment. Obviously consumption-goods, taken as a whole, have in this sense the longest period of production, since of every productive process they constitute the last stage. Thus if the first impulse towards the increase in effective demand comes from an increase in consumption, the initial elasticity of employment will be further below its eventual equilibrium-level than if the impulse comes from an increase in investment. Moreover, if the increased demand is directed to products with a relatively low elasticity of employment, a larger proportion of it will go to swell the incomes of entrepreneurs and a smaller proportion to swell the incomes of wage-earners and other prime-cost factors; with the possible result that the repercussions may be somewhat less favourable to expenditure, owing to the likelihood of entrepreneurs saving more of their increment of income than wage-earners would. Nevertheless the distinction between the two cases must not be over-stated, since a large part of the reactions will be much the same in both.[2]

---

[1] This is not identical with the usual definition, but it seems to me to embody what is significant in the idea.

[2] Some further discussion of the above topic is to be found in my *Treatise on Money* [*JMK*, vol. v], Book IV.

However long the notice given to entrepreneurs of a prospective change in demand, it is not possible for the initial elasticity of employment, in response to a *given* increase of investment, to be as great as its eventual equilibrium value, unless there are surplus stocks and surplus capacity at every stage of production. On the other hand, the depletion of the surplus stocks will have an offsetting effect on the amount by which investment increases. If we suppose that there are initially some surpluses at every point, the initial elasticity of employment may approximate to unity; then after the stocks have been absorbed, but before an increased supply is coming forward at an adequate rate from the earlier stages of production, the elasticity will fall away; rising again towards unity as the new position of equilibrium is approached. This is subject, however, to some qualification in so far as there are rent factors which absorb more expenditure as employment increases, or if the rate of interest increases. For these reasons perfect stability of prices is impossible in an economy subject to change—unless, indeed, there is some peculiar mechanism which ensures temporary fluctuations of just the right degree in the propensity to consume. But price-instability arising in this way does not lead to the kind of profit stimulus which is liable to bring into existence excess capacity. For the windfall gain will wholly accrue to those entrepreneurs who happen to possess products at a relatively advanced stage of production, and there is nothing which the entrepreneur, who does not possess specialised resources of the right kind, can do to attract this gain to himself. Thus the inevitable price-instability due to change cannot affect the *actions* of entrepreneurs, but merely directs a *de facto* windfall of wealth into the laps of the lucky ones (*mutatis mutandis* when the supposed change is in the other direction). This fact has, I think, been overlooked in some contemporary discussions of a practical policy aimed at stabilising prices.

It is true that in a society liable to change such a policy cannot be perfectly successful. But it does not follow that every small temporary departure from price stability necessarily sets up a cumulative disequilibrium.

III

We have shown that when effective demand is deficient there is under-employment of labour in the sense that there are men unemployed who would be willing to work at less than the existing real wage. Consequently, as effective demand increases, employment increases, though at a real wage equal to or less than the existing one, until a point comes at which there is no surplus of labour available at the then existing real wage; i.e. no more men (or hours of labour) available unless money-wages rise (from this point onwards) *faster* than prices. The next problem is to consider what will happen if, when this point has been reached, expenditure still continues to increase.

Up to this point the decreasing return from applying more labour to a given capital equipment has been offset by the acquiescence of labour in a diminishing real wage. But after this point a unit of labour would require the inducement of the equivalent of an increased quantity of product, whereas the yield from applying a further unit would be a diminished quantity of product. The conditions of strict equilibrium require, therefore, that wages and prices, and consequently profits also, should all rise in the same proportion as expenditure, the 'real' position, including the volume of output and employment, being left unchanged in all respects. We have reached, that is to say, a situation in which the crude quantity theory of money (interpreting 'velocity' to mean 'income-velocity') is fully satisfied; for output does not alter and prices rise in exact proportion to MV.

Nevertheless there are certain practical qualifications to this conclusion which must be borne in mind in applying it to an actual case:

(1) For a time at least, rising prices may delude entrepreneurs into increasing employment beyond the level which maximises their individual profits measured in terms of the product. For they are so accustomed to regard rising sale-proceeds in terms of money as a signal for expanding production, that they may continue to do so when this policy has in fact ceased to be to their best advantage; i.e. they may underestimate their marginal user cost in the new price environment.

(2) Since that part of his profit which the entrepreneur has to hand on to the rentier is fixed in terms of money, rising prices, even though unaccompanied by any change in output, will re-distribute incomes to the advantage of the entrepreneur and to the disadvantage of the rentier, which may have a reaction on the propensity to consume. This, however, is not a process which will have only begun when full employment has been attained;—it will have been making steady progress all the time that the expenditure was increasing. If the rentier is less prone to spend than the entrepreneur, the gradual withdrawal of real income from the former will mean that full employment will be reached with a smaller increase in the quantity of money and a smaller reduction in the rate of interest than will be the case if the opposite hypothesis holds. After full employment has been reached, a further rise of prices will, if the first hypothesis continues to hold, mean that the rate of interest will have to rise somewhat to prevent prices from rising indefinitely, and that the increase in the quantity of money will be less than in proportion to the increase in expenditure; whilst if the second hypothesis holds, the opposite will be the case. It may be that, as the real income of the rentier is diminished, a point will come when, as a result of his growing relative impoverishment, there will be a change-over from the first hypothesis to the second, which

point may be reached either before or after full employment has been attained.

IV

There is, perhaps, something a little perplexing in the apparent asymmetry between inflation and deflation. For whilst a deflation of effective demand below the level required for full employment will diminish employment as well as prices, an inflation of it above this level will merely affect prices. This asymmetry is, however, merely a reflection of the fact that, whilst labour is always in a position to refuse to work on a scale involving a real wage which is less than the marginal disutility of that amount of employment, it is not in a position to insist on being offered work on a scale involving a real wage which is not greater than the marginal disutility of that amount of employment.

Chapter 21

THE THEORY OF PRICES

I

So long as economists are concerned with what is called
the theory of value, they have been accustomed to
teach that prices are governed by the conditions of
supply and demand; and, in particular, changes in
marginal cost and the elasticity of short-period supply
have played a prominent part. But when they pass in
volume II, or more often in a separate treatise, to the
theory of money and prices, we hear no more of these
homely but intelligible concepts and move into a world
where prices are governed by the quantity of money,
by its income-velocity, by the velocity of circulation
relatively to the volume of transactions, by hoarding,
by forced saving, by inflation and deflation *et hoc genus
omne*; and little or no attempt is made to relate these
vaguer phrases to our former notions of the elasticities
of supply and demand. If we reflect on what we are
being taught and try to rationalise it, in the simpler
discussions it seems that the elasticity of supply must
have become zero and demand proportional to the
quantity of money; whilst in the more sophisticated
we are lost in a haze where nothing is clear and everything is possible. We have all of us become used to
finding ourselves sometimes on the one side of the moon
and sometimes on the other, without knowing what
route or journey connects them, related, apparently,
after the fashion of our waking and our dreaming lives.

One of the objects of the foregoing chapters has been to escape from this double life and to bring the theory of prices as a whole back to close contact with the theory of value. The division of economics between the theory of value and distribution on the one hand and the theory of money on the other hand is, I think, a false division. The right dichotomy is, I suggest, between the theory of the individual industry or firm and of the rewards and the distribution between different uses of a *given* quantity of resources on the one hand, and the theory of output and employment *as a whole* on the other hand. So long as we limit ourselves to the study of the individual industry or firm on the assumption that the aggregate quantity of employed resources is constant, and, provisionally, that the conditions of other industries or firms are unchanged, it is true that we are not concerned with the significant characteristics of money. But as soon as we pass to the problem of what determines output and employment as a whole, we require the complete theory of a monetary economy.

Or, perhaps, we might make our line of division between the theory of stationary equilibrium and the theory of shifting equilibrium—meaning by the latter the theory of a system in which changing views about the future are capable of influencing the present situation. *For the importance of money essentially flows from its being a link between the present and the future.* We can consider what distribution of resources between different uses will be consistent with equilibrium under the influence of normal economic motives in a world in which our views concerning the future are fixed and reliable in all respects;—with a further division, perhaps, between an economy which is unchanging and one subject to change, but where all things are foreseen from the beginning. Or we can pass from this simplified propaedeutic to the problems of the real world in which our previous expectations are liable to disappointment and expectations concerning the future affect what we do to-day. It is when we have made this transition that the peculiar properties of money as a link between the present and the future must enter into our calculations. But, although the theory of shifting equilibrium must necessarily be pursued in terms of a monetary economy, it remains a theory of value and distribution and not a separate 'theory of money'. Money in its significant attributes is, above all, a subtle device for linking the present to the future; and we cannot even begin to discuss the effect of changing expectations on current activities except in monetary terms. We cannot get rid of money even by abolishing gold and silver and legal tender instruments. So long as there exists any durable asset, it is capable of possessing monetary attributes[1] and, therefore, of giving rise to the characteristic problems of a monetary economy.

II

In a single industry its particular price-level depends partly on the rate of remuneration of the factors of production which enter into its marginal cost, and partly on the scale of output. There is no reason to modify this conclusion when we pass to industry as a whole. The general price-level depends partly on the rate of remuneration of the factors of production which enter into marginal cost and partly on the scale of output as a whole, i.e. (taking equipment and technique as given) on the volume of employment. It is true that, when we pass to output as a whole, the costs of production in any industry partly depend on the output of other industries. But the more significant change, of which we have to take account, is the effect of changes in *demand* both on costs and on volume. It is on the side of demand that we have to introduce quite new ideas when we are dealing with demand as a whole and

[1] Cf. chapter 17 above.

no longer with the demand for a single product taken in isolation, with demand as a whole assumed to be unchanged.

III

If we allow ourselves the simplification of assuming that the rates of remuneration of the different factors of production which enter into marginal cost all change in the same proportion, i.e. in the same proportion as the wage-unit, it follows that the general price-level (taking equipment and technique as given) depends partly on the wage-unit and partly on the volume of employment. Hence the effect of changes in the quantity of money on the price-level can be considered as being compounded of the effect on the wage-unit and the effect on employment.

To elucidate the ideas involved, let us simplify our assumptions still further, and assume (1) that all unemployed resources are homogeneous and interchangeable in their efficiency to produce what is wanted, and (2) that the factors of production entering into marginal cost are content with the same money-wage so long as there is a surplus of them unemployed. In this case we have constant returns and a rigid wage-unit, so long as there is any unemployment. It follows that an increase in the quantity of money will have no effect whatever on prices, so long as there is any unemployment, and that employment will increase in exact proportion to any increase in effective demand brought about by the increase in the quantity of money; whilst as soon as full employment is reached, it will thenceforward be the wage-unit and prices which will increase in exact proportion to the increase in effective demand. Thus if there is perfectly elastic supply so long as there is unemployment, and perfectly inelastic supply so soon as full employment is reached, and if effective demand changes in the same proportion as the quantity of money, the quantity theory of money can be enunciated as follows: 'So long as there is unemployment, *employment* will change in the same proportion as the quantity of money; and when there is full employment, *prices* will change in the same proportion as the quantity of money'.

Having, however, satisfied tradition by introducing a sufficient number of simplifying assumptions to enable us to enunciate a quantity theory of money, let us now consider the possible complications which will in fact influence events:

(1) Effective demand will not change in exact proportion to the quantity of money.

(2) Since resources are not homogeneous, there will be diminishing, and not constant, returns as employment gradually increases.

(3) Since resources are not interchangeable, some commodities will reach a condition of inelastic supply whilst there are still unemployed resources available for the production of other commodities.

(4) The wage-unit will tend to rise, before full employment has been reached.

(5) The remunerations of the factors entering into marginal cost will not all change in the same proportion.

Thus we must first consider the effect of changes in the quantity of money on the quantity of effective demand; and the increase in effective demand will, generally speaking, spend itself partly in increasing the quantity of employment and partly in raising the level of prices. Thus instead of constant prices in conditions of unemployment, and of prices rising in proportion to the quantity of money in conditions of full employment, we have in fact a condition of prices rising gradually as employment increases. The theory of prices, that is to say, the analysis of the relation between changes in the quantity of money and changes in the price-level with a view to determining the elasticity of prices in response to changes in the quantity of money, must, therefore, direct itself to the five complicating factors set forth above.

We will consider each of them in turn. But this procedure must not be allowed to lead us into supposing that they are, strictly speaking, independent. For example, the proportion, in which an increase in effective demand is divided in its effect between increasing output and raising prices, may affect the way in which the quantity of money is related to the quantity of effective demand. Or, again, the differences in the proportions, in which the remunerations of different factors change, may influence the relation between the quantity of money and the quantity of effective demand. The object of our analysis is, not to provide a machine, or method of blind manipulation, which will furnish an infallible answer, but to provide ourselves with an organised and orderly method of thinking out particular problems; and, after we have reached a provisional conclusion by isolating the complicating factors one by one, we then have to go back on ourselves and allow, as well as we can, for the probable interactions of the factors amongst themselves. This is the nature of economic thinking. Any other way of applying our formal principles of thought (without which, however, we shall be lost in the wood) will lead us into error. It is a great fault of symbolic pseudo-mathematical methods of formalising a system of economic analysis, such as we shall set down in section VI of this chapter, that they expressly assume strict independence between the factors involved and lose all their cogency and authority if this hypothesis is disallowed; whereas, in ordinary discourse, where we are not blindly manipulating but know all the time what we are doing and what the words mean, we can keep 'at the back of our heads' the necessary reserves and qualifications and the adjustments which we shall have to make later on, in a way in which we cannot keep complicated partial differentials 'at the back' of several pages of algebra which assume that they all vanish. Too large a proportion of recent 'mathematical' economics are merely concoctions, as imprecise as the initial assumptions they rest on, which allow the author to lose sight of the complexities and interdependencies of the real world in a maze of pretentious and unhelpful symbols.

IV

(1) The primary effect of a change in the quantity of money on the quantity of effective demand is through its influence on the rate of interest. If this were the only reaction, the quantitative effect could be derived from the three elements—(a) the schedule of liquidity-preference which tells us by how much the rate of interest will have to fall in order that the new money may be absorbed by willing holders, (b) the schedule of marginal efficiencies which tells us by how much a given fall in the rate of interest will increase investment, and (c) the investment multiplier which tells us by how much a given increase in investment will increase effective demand as a whole.

But this analysis, though it is valuable in introducing order and method into our enquiry, presents a deceptive simplicity, if we forget that the three elements (a), (b) and (c) are themselves partly dependent on the complicating factors (2), (3), (4) and (5) which we have not yet considered. For the schedule of liquidity-preference itself depends on how much of the new money is absorbed into the income and industrial circulations, which depends in turn on how much effective demand increases and how the increase is divided between the rise of prices, the rise of wages, and the volume of output and employment. Furthermore, the schedule of marginal efficiencies will partly depend on the effect which the circumstances attendant on the increase in the quantity of money have on expectations of the future monetary prospects. And finally the multiplier will be influenced by the way in which the new income resulting from the increased effective demand is distributed between different classes of consumers. Nor, of course, is this list of possible interactions complete. Nevertheless, if we have all the facts before us, we shall have enough simultaneous equations to give us a determinate result. There will be a determinate amount of increase in the quantity of effective demand which, after taking everything into account, will correspond to, and be in equilibrium with, the increase in the quantity of money. Moreover, it is only in highly exceptional circumstances that an increase in the quantity of money will be associated with a *decrease* in the quantity of effective demand.

The ratio between the quantity of effective demand and the quantity of money closely corresponds to what is often called the 'income-velocity of money';— except that effective demand corresponds to the income the expectation of which has set production moving, not to the actually realised income, and to gross, not net, income. But the 'income-velocity of money' is, in itself, merely a name which explains nothing. There is no reason to expect that it will be constant. For it depends, as the foregoing discussion has shown, on many complex and variable factors. The use of this term obscures, I think, the real character of the causation, and has led to nothing but confusion.

(2) As we have shown above (p. 42), the distinction between diminishing and constant returns partly depends on whether workers are remunerated in strict proportion to their efficiency. If so, we shall have constant labour-costs (in terms of the wage-unit) when employment increases. But if the wage of a given grade of labourers is uniform irrespective of the efficiency of the individuals, we shall have rising labour-costs, irrespective of the efficiency of the equipment. Moreover, if equipment is non-homogeneous and some part of it involves a greater prime cost per

unit of output, we shall have increasing marginal prime costs over and above any increase due to increasing labour-costs.

Hence, in general, supply price will increase as output from a given equipment is increased. Thus increasing output will be associated with rising prices, apart from any change in the wage-unit.

(3) Under (2) we have been contemplating the possibility of supply being imperfectly elastic. If there is a perfect balance in the respective quantities of specialised unemployed resources, the point of full employment will be reached for all of them simultaneously. But, in general, the demand for some services and commodities will reach a level beyond which their supply is, for the time being, perfectly inelastic, whilst in other directions there is still a substantial surplus of resources without employment. Thus as output increases, a series of 'bottle-necks' will be successively reached, where the supply of particular commodities ceases to be elastic and their prices have to rise to whatever level is necessary to divert demand into other directions.

It is probable that the general level of prices will not rise very much as output increases, so long as there are available efficient unemployed resources of every type. But as soon as output has increased sufficiently to begin to reach the 'bottle-necks', there is likely to be a sharp rise in the prices of certain commodities.

Under this heading, however, as also under heading (2), the elasticity of supply partly depends on the elapse of time. If we assume a sufficient interval for the quantity of equipment itself to change, the elasticities of supply will be decidedly greater eventually. Thus a moderate change in effective demand, coming on a situation where there is widespread unemployment, may spend itself very little in raising prices and mainly in increasing employment; whilst a larger change, which, being unforeseen, causes some temporary 'bottle-necks' to be reached, will spend itself in raising prices, as distinct from employment, to a greater extent at first than subsequently.

(4) That the wage-unit may tend to rise before full employment has been reached, requires little comment or explanation. Since each group of workers will gain, *cet. par.*, by a rise in its own wages, there is naturally for all groups a pressure in this direction, which entrepreneurs will be more ready to meet when they are doing better business. For this reason a proportion of any increase in effective demand is likely to be absorbed in satisfying the upward tendency of the wage-unit.

Thus, in addition to the final critical point of full employment at which money-wages have to rise, in response to an increasing effective demand in terms of money, fully in proportion to the rise in the prices of wage-goods, we have a succession of earlier semi-critical points at which an increasing effective demand tends to raise money-wages though not fully in proportion to the rise in the price of wage-goods; and similarly in the case of a decreasing effective demand. In actual experience the wage-unit does not change continuously in terms of money in response to every small change in effective demand; but discontinuously. These points of discontinuity are determined by the psychology of the workers and by the policies of employers and trade unions. In an open system, where they mean a change relatively to wage-costs elsewhere, and in a trade cycle, where even in a closed system they may mean a change relatively to expected wage-costs in the future, they can be of considerable practical significance. These points, where a further increase in effective demand in terms of money is liable to cause a discontinuous rise in the wage-unit, might be deemed, from a certain point of view, to be positions of semi-inflation, having some analogy (though a very imperfect one) to the absolute inflation (cf. p. 303 below) which ensues on an increase in effective demand in circumstances of full employment. They have, moreover, a good deal of historical importance. But they do not readily lend themselves to theoretical generalisations.

(5) Our first simplification consisted in assuming that the remunerations of the various factors entering into marginal cost all change in the same proportion. But in fact the rates of remuneration of different factors in terms of money will show varying degrees of rigidity and they may also have different elasticities of supply in response to changes in the money-rewards offered. If it were not for this, we could say that the price-level is compounded of two factors, the wage-unit and the quantity of employment.

Perhaps the most important element in marginal cost which is likely to change in a different proportion from the wage-unit, and also to fluctuate within much wider limits, is marginal user cost. For marginal user cost may increase sharply when employment begins to improve, if (as will probably be the case) the increasing effective demand brings a rapid change in the prevailing expectation as to the date when the replacement of equipment will be necessary.

Whilst it is for many purposes a very useful first approximation to assume that the rewards of all the factors entering into marginal prime-cost change in the same proportion as the wage-unit, it might be better, perhaps, to take a weighted average of the rewards of the factors entering into marginal prime-cost, and call this the *cost-unit*. The cost-unit, or, subject to the above approximation, the wage-unit, can thus be regarded as the essential standard of value; and the price-level, given the state of technique and equipment, will depend partly on the cost-unit and partly on the scale of output, increasing, where output increases, *more* than in proportion to any increase in the cost-unit, in accordance with the principle of diminishing returns in the short period. We have full employment when output has risen to a level at which the marginal return from a representative unit of the factors of production has fallen to the minimum figure at which a quantity of the factors sufficient to produce this output is available.

V

When a further increase in the quantity of effective demand produces no further increase in output and entirely spends itself on an increase in the cost-unit fully proportionate to the increase in effective demand, we have reached a condition which might be appropriately designated as one of true inflation. Up to this point the effect of monetary expansion is entirely a question of degree, and there is no previous point at which we can draw a definite line and declare that conditions of inflation have set in. Every previous increase in the quantity of money is likely, in so far as it increases effective demand, to spend itself partly in increasing the cost-unit and partly in increasing output.

It appears, therefore, that we have a sort of asymmetry on the two sides of the critical level above which true inflation sets in. For a contraction of effective demand below the critical level will reduce its amount measured in cost-units; whereas an expansion of effective demand beyond this level will not, in general, have the effect of increasing its amount in terms of cost-units. This result follows from the assumption that the factors of production, and in particular the workers, are disposed to resist a reduction in their money-rewards, and that there is no corresponding motive to resist an increase. This assumption is, however, obviously well founded in the facts, due to the circumstance that a change, which is not an all-round change, is beneficial to the special factors affected when it is upward and harmful when it is downward.

If, on the contrary, money-wages were to fall with-

303

out limit whenever there was a tendency for less than full employment, the asymmetry would, indeed, disappear. But in that case there would be no resting-place below full employment until either the rate of interest was incapable of falling further or wages were zero. In fact we must have *some* factor, the value of which in terms of money is, if not fixed, at least sticky, to give us any stability of values in a monetary system.

The view that *any* increase in the quantity of money is inflationary (unless we mean by *inflationary* merely that prices are rising) is bound up with the underlying assumption of the classical theory that we are *always* in a condition where a reduction in the real rewards of the factors of production will lead to a curtailment in their supply.

VI

With the aid of the notation introduced in chapter 20 we can, if we wish, express the substance of the above in symbolic form.

Let us write $MV = D$ where M is the quantity of money, V its income-velocity (this definition differing in the minor respects indicated above from the usual definition) and D the effective demand. If, then, V is constant, prices will change in the same proportion as the quantity of money provided that $e_p\left(= \dfrac{Ddp}{pdD}\right)$ is unity. This condition is satisfied (see p. 286 above) if $e_o = 0$ or if $e_w = 1$. The condition $e_w = 1$ means that the wage-unit in terms of money rises in the same proportion as the effective demand, since $e_w = \dfrac{DdW}{WdD}$; and the condition $e_o = 0$ means that output no longer shows any response to a further increase in effective demand, since $e_o = \dfrac{DdO}{OdD}$. Output in either case will be unaltered.

Next, we can deal with the case where income-velocity is not constant, by introducing yet a further

304

elasticity, namely the elasticity of effective demand in response to changes in the quantity of money,

$$e_d = \frac{MdD}{DdM}.$$

This gives us

$$\frac{Mdp}{pdM} = e_p . e_d \text{ where } e_p = 1 - e_e . e_o (1 - e_w);$$

so that
$$e = e_d - (1 - e_w) e_d . e_e e_o$$
$$= e_d (1 - e_e e_o + e_e e_o . e_w)$$

where e without suffix $\left(= \dfrac{Mdp}{pdM} \right)$ stands for the apex of this pyramid and measures the response of money-prices to changes in the quantity of money.

Since this last expression gives us the proportionate change in prices in response to a change in the quantity of money, it can be regarded as a generalised statement of the quantity theory of money. I do not myself attach much value to manipulations of this kind; and I would repeat the warning, which I have given above, that they involve just as much tacit assumption as to what variables are taken as independent (partial differentials being ignored throughout) as does ordinary discourse, whilst I doubt if they carry us any further than ordinary discourse can. Perhaps the best purpose served by writing them down is to exhibit the extreme complexity of the relationship between prices and the quantity of money, when we attempt to express it in a formal manner. It is, however, worth pointing out that, of the four terms e_d, e_w, e_e and e_o upon which the effect on prices of changes in the quantity of money depends, e_d stands for the liquidity factors which determine the demand for money in each situation, e_w for the labour factors (or, more strictly, the factors entering into prime-cost) which determine the extent to which money-wages are raised as employment increases, and e_e and e_o for the physical factors which determine the rate of decreasing returns as more employment is applied to the existing equipment.

If the public hold a constant proportion of their income in money, $e_d = 1$; if money-wages are fixed, $e_w = 0$; if there are constant returns throughout so that marginal return equals average return, $e_e e_o = 1$; and if there is full employment either of labour or of equipment, $e_e e_o = 0$.

Now $e = 1$, if $e_d = 1$, and $e_w = 1$; or if $e_d = 1$, $e_w = 0$ and $e_e . e_o = 0$; or if $e_d = 1$ and $e_o = 0$. And obviously there is a variety of other special cases in which $e = 1$. But in general e is not unity; and it is, perhaps, safe to make the generalisation that on plausible assumptions relating to the real world, and excluding the case of a 'flight from the currency' in which e_d and e_w become large, e is, as a rule, less than unity.

VII

So far, we have been primarily concerned with the way in which changes in the quantity of money affect prices in the short period. But in the long run is there not some simpler relationship?

This is a question for historical generalisation rather than for pure theory. If there is some tendency to a measure of long-run uniformity in the state of liquidity-preference, there may well be some sort of rough relationship between the national income and the quantity of money required to satisfy liquidity-preference, taken as a mean over periods of pessimism and optimism together. There may be, for example, some fairly stable proportion of the national income more than which people will not readily keep in the shape of idle balances for long periods together, provided the rate of interest exceeds a certain psychological minimum; so that if the quantity of money beyond what is required in the active circulation is in excess of this proportion of the national income, there will be a tendency sooner or later for the rate of interest to fall to the neighbourhood of this minimum. The falling rate of interest will then, *cet. par.*, increase effective demand, and the increasing effective demand will reach one or more of the semi-critical points at which the wage-unit will tend to show a discontinuous rise, with a corresponding effect on prices. The opposite tendencies will set in if the quantity of surplus money is an abnormally low proportion of the national income. Thus the net effect of fluctuations over a period of time will be to establish a mean figure in conformity with the stable proportion between the national income and the quantity of money to which the psychology of the public tends sooner or later to revert.

These tendencies will probably work with less friction in the upward than in the downward direction. But if the quantity of money remains very deficient for a long time, the escape will be normally found in changing the monetary standard or the monetary system so as to raise the quantity of money, rather than in forcing down the wage-unit and thereby increasing the burden of debt. Thus the very long-run course of prices has almost always been upward. For when money is relatively abundant, the wage-unit rises; and when money is relatively scarce, some means is found to increase the effective quantity of money.

During the nineteenth century, the growth of population and of invention, the opening-up of new lands, the state of confidence and the frequency of war over the average of (say) each decade seem to have been sufficient, taken in conjunction with the propensity to consume, to establish a schedule of the marginal efficiency of capital which allowed a reasonably satisfactory average level of employment to be compatible with a rate of interest high enough to be psychologically acceptable to wealth-owners. There is evidence that for a period of almost one hundred and fifty years the long-run typical rate of interest in the leading financial centres was about 5 per cent, and the gilt-edged rate between 3 and $3\frac{1}{2}$ per cent; and that these rates of interest were modest enough to encourage a rate of investment consistent with an average of employment which was not intolerably low. Sometimes the wage-unit, but more often the monetary standard or the monetary system (in particular through the development of bank-money), would be adjusted so as to ensure that the quantity of money in terms of wage-units was sufficient to satisfy normal liquidity-preference at rates of interest which were seldom much below the standard rates indicated above. The tendency of the wage-unit was, as usual, steadily upwards on the whole, but the efficiency of labour was also increasing. Thus the balance of forces was such as to allow a fair measure of stability of prices;—the highest quinquennial average for Sauerbeck's index number between 1820 and 1914 was only 50 per cent above the lowest. This was not accidental. It is rightly described as due to a balance of forces in an age when individual groups of employers were strong enough to prevent the wage-unit from rising much faster than the efficiency of production, and when monetary systems were at the same time sufficiently fluid and sufficiently conservative to provide an average supply of money in terms of wage-units which allowed to prevail the lowest average rate of interest readily acceptable by wealth-owners under the influence of their liquidity-preferences. The average level of employment was, of course, substantially below full employment, but not so intolerably below it as to provoke revolutionary changes.

To-day, and presumably for the future, the schedule of the marginal efficiency of capital is, for a variety of reasons, much lower than it was in the nineteenth century. The acuteness and the peculiarity of our contemporary problem arises, therefore, out of the possibility that the average rate of interest which will allow a reasonable average level of employment is one so unacceptable to wealth-owners that it cannot be readily established merely by manipulating the quantity of money. So long as a tolerable level of employment could be attained on the average of one or two or three decades merely by assuring an adequate supply of money in terms of wage-units, even the nineteenth century could find a way. If this was our only problem now—if a sufficient degree of devaluation is all we need—we, to-day, would certainly find a way.

But the most stable, and the least easily shifted, element in our contemporary economy has been hitherto, and may prove to be in future, the minimum rate of interest acceptable to the generality of wealth-owners.[1] If a tolerable level of employment requires a rate of interest much below the average rates which ruled in the nineteenth century, it is most doubtful whether it can be achieved merely by manipulating the quantity of money. From the percentage gain, which the schedule of marginal efficiency of capital allows the borrower to expect to earn, there has to be deducted (1) the cost of bringing borrowers and lenders together, (2) income and sur-taxes and (3) the allowance which the lender requires to cover his risk and uncertainty, before we arrive at the net yield available to tempt the wealth-owner to sacrifice his liquidity. If, in conditions of tolerable average employment, this net yield turns out to be infinitesimal, time-honoured methods may prove unavailing.

To return to our immediate subject, the long-run relationship between the national income and the quantity of money will depend on liquidity-preferences. And the long-run stability or instability of prices will depend on the strength of the upward trend of the wage-unit (or, more precisely, of the cost-unit) compared with the rate of increase in the efficiency of the productive system.

[1] Cf. the nineteenth-century saying, quoted by Bagehot, that 'John Bull can stand many things, but he cannot stand 2 per cent.'

BOOK VI

SHORT NOTES SUGGESTED BY THE GENERAL THEORY

Chapter 22

NOTES ON THE TRADE CYCLE

Since we claim to have shown in the preceding chapters what determines the volume of employment at any time, it follows, if we are right, that our theory must be capable of explaining the phenomena of the trade cycle.

If we examine the details of any actual instance of the trade cycle, we shall find that it is highly complex and that every element in our analysis will be required for its complete explanation. In particular we shall find that fluctuations in the propensity to consume, in the state of liquidity-preference, and in the marginal efficiency of capital have all played a part. But I suggest that the essential character of the trade cycle and, especially, the regularity of time-sequence and of duration which justifies us in calling it a *cycle*, is mainly due to the way in which the marginal efficiency of capital fluctuates. The trade cycle is best regarded, I think, as being occasioned by a cyclical change in the marginal efficiency of capital, though complicated and often aggravated by associated changes in the other significant short-period variables of the economic system. To develop this thesis would occupy a book rather than a chapter, and would require a close examination of facts. But the following short notes will be sufficient to indicate the line of investigation which our preceding theory suggests.

I

By a *cyclical* movement we mean that as the system progresses in, e.g. the upward direction, the forces

313

propelling it upwards at first gather force and have a cumulative effect on one another but gradually lose their strength until at a certain point they tend to be replaced by forces operating in the opposite direction; which in turn gather force for a time and accentuate one another, until they too, having reached their maximum development, wane and give place to their opposite. We do not, however, merely mean by a *cyclical* movement that upward and downward tendencies, once started, do not persist for ever in the same direction but are ultimately reversed. We mean also that there is some recognisable degree of regularity in the time-sequence and duration of the upward and downward movements.

There is, however, another characteristic of what we call the trade cycle which our explanation must cover if it is to be adequate; namely, the phenomenon of the *crisis*—the fact that the substitution of a downward for an upward tendency often takes place suddenly and violently, whereas there is, as a rule, no such sharp turning-point when an upward is substituted for a downward tendency.

Any fluctuation in investment not offset by a corresponding change in the propensity to consume will, of course, result in a fluctuation in employment. Since, therefore, the volume of investment is subject to highly complex influences, it is highly improbable that all fluctuations either in investment itself or in the marginal efficiency of capital will be of a cyclical character. One special case, in particular, namely, that which is associated with agricultural fluctuations, will be separately considered in a later section of this chapter. I suggest, however, that there are certain definite reasons why, in the case of a typical industrial trade cycle in the nineteenth-century environment, fluctuations in the marginal efficiency of capital should have had cyclical characteristics. These reasons are by no means unfamiliar either in themselves or as explanations of the trade cycle. My only purpose here is to link them up with the preceding theory.

II

I can best introduce what I have to say by beginning with the later stages of the boom and the onset of the 'crisis'.

We have seen above that the marginal efficiency of capital[1] depends, not only on the existing abundance or scarcity of capital-goods and the current cost of production of capital-goods, but also on current expectations as to the future yield of capital-goods. In the case of durable assets it is, therefore, natural and reasonable that expectations of the future should play a dominant part in determining the scale on which new investment is deemed advisable. But, as we have seen, the basis for such expectations is very precarious. Being based on shifting and unreliable evidence, they are subject to sudden and violent changes.

Now, we have been accustomed in explaining the 'crisis' to lay stress on the rising tendency of the rate of interest under the influence of the increased demand for money both for trade and speculative purposes. At times this factor may certainly play an aggravating and, occasionally perhaps, an initiating part. But I suggest that a more typical, and often the predominant, explanation of the crisis is, not primarily a rise in the rate of interest, but a sudden collapse in the marginal efficiency of capital.

The later stages of the boom are characterised by optimistic expectations as to the future yield of capital-goods sufficiently strong to offset their growing abundance and their rising costs of production and, probably, a rise in the rate of interest also. It is of the nature of

---

[1] It is often convenient in contexts where there is no room for misunderstanding to write 'the marginal efficiency of capital', where 'the schedule of the marginal efficiency of capital' is meant.

organised investment markets, under the influence of purchasers largely ignorant of what they are buying and of speculators who are more concerned with forecasting the next shift of market sentiment than with a reasonable estimate of the future yield of capital-assets, that, when disillusion falls upon an over-optimistic and over-bought market, it should fall with sudden and even catastrophic force.[1] Moreover, the dismay and uncertainty as to the future which accompanies a collapse in the marginal efficiency of capital naturally precipitates a sharp increase in liquidity-preference—and hence a rise in the rate of interest. Thus the fact that a collapse in the marginal efficiency of capital tends to be associated with a rise in the rate of interest may seriously aggravate the decline in investment. But the essence of the situation is to be found, nevertheless, in the collapse in the marginal efficiency of capital, particularly in the case of those types of capital which have been contributing most to the previous phase of heavy new investment. Liquidity-preference, except those manifestations of it which are associated with increasing trade and speculation, does not increase until *after* the collapse in the marginal efficiency of capital.

It is this, indeed, which renders the slump so intractable. Later on, a decline in the rate of interest will be a great aid to recovery and, probably, a necessary condition of it. But, for the moment, the collapse in the marginal efficiency of capital may be so complete that no practicable reduction in the rate of interest will be enough. If a reduction in the rate of interest was capable of proving an effective remedy by itself, it might be possible to achieve a recovery without the elapse of any considerable interval of time and by means more or less directly under the control of the monetary

---

[1] I have shown above (chapter 12) that, although the private investor is seldom himself directly responsible for new investment, nevertheless the entrepreneurs, who are directly responsible, will find it financially advantageous, and often unavoidable, to fall in with the ideas of the market, even though they themselves are better instructed.

authority. But, in fact, this is not usually the case; and it is not so easy to revive the marginal efficiency of capital, determined, as it is, by the uncontrollable and disobedient psychology of the business world. It is the return of confidence, to speak in ordinary language, which is so insusceptible to control in an economy of individualistic capitalism. This is the aspect of the slump which bankers and business men have been right in emphasising, and which the economists who have put their faith in a 'purely monetary' remedy have underestimated.

This brings me to my point. The explanation of the *time-element* in the trade cycle, of the fact that an interval of time of a particular order of magnitude must usually elapse before recovery begins, is to be sought in the influences which govern the recovery of the marginal efficiency of capital. There are reasons, given firstly by the length of life of durable assets in relation to the normal rate of growth in a given epoch, and secondly by the carrying-costs of surplus stocks, why the duration of the downward movement should have an order of magnitude which is not fortuitous, which does not fluctuate between, say, one year this time and ten years next time, but which shows some regularity of habit between, let us say, three and five years.

Let us recur to what happens at the crisis. So long as the boom was continuing, much of the new investment showed a not unsatisfactory current yield. The disillusion comes because doubts suddenly arise concerning the reliability of the prospective yield, perhaps because the current yield shows signs of falling off, as the stock of newly produced durable goods steadily increases. If current costs of production are thought to be higher than they will be later on, that will be a further reason for a fall in the marginal efficiency of capital. Once doubt begins it spreads rapidly. Thus at the outset of the slump there is probably much capital of which the marginal efficiency has become negligible or even negative. But the interval of time, which will have to elapse before the shortage of capital through use, decay and obsolescence causes a sufficiently obvious scarcity to increase the marginal efficiency, may be a somewhat stable function of the average durability of capital in a given epoch. If the characteristics of the epoch shift, the standard time-interval will change. If, for example, we pass from a period of increasing population into one of declining population, the characteristic phase of the cycle will be lengthened. But we have in the above a substantial reason why the duration of the slump should have a definite relationship to the length of life of durable assets and to the normal rate of growth in a given epoch.

The second stable time-factor is due to the carrying-costs of surplus stocks which force their absorption within a certain period, neither very short nor very long. The sudden cessation of new investment after the crisis will probably lead to an accumulation of surplus stocks of unfinished goods. The carrying-costs of these stocks will seldom be less than 10 per cent. per annum. Thus the fall in their price needs to be sufficient to bring about a restriction which provides for their absorption within a period of, say, three to five years at the outside. Now the process of absorbing the stocks represents negative investment, which is a further deterrent to employment; and, when it is over, a manifest relief will be experienced.

Moreover, the reduction in working capital, which is necessarily attendant on the decline in output on the downward phase, represents a further element of dis-investment, which may be large; and, once the recession has begun, this exerts a strong cumulative influence in the downward direction. In the earliest phase of a typical slump there will probably be an investment in increasing stocks which helps to offset disinvestment in working-capital; in the next phase there may be a short period of disinvestment both in stocks and in working-capital; after the lowest point has been passed there is likely to be a further disinvestment in stocks which partially offsets reinvestment in working-capital; and, finally, after the recovery is well on its way, both factors will be simultaneously favourable to investment. It is against this background that the additional and super-imposed effects of fluctuations of investment in durable goods must be examined. When a decline in this type of investment has set a cyclical fluctuation in motion there will be little encouragement to a recovery in such investment until the cycle has partly run its course.[1]

Unfortunately a serious fall in the marginal efficiency of capital also tends to affect adversely the propensity to consume. For it involves a severe decline in the market value of stock exchange equities. Now, on the class who take an active interest in their stock exchange investments, especially if they are em-ploying borrowed funds, this naturally exerts a very depressing influence. These people are, perhaps, even more influenced in their readiness to spend by rises and falls in the value of their investments than by the state of their incomes. With a 'stock-minded' public as in the United States to-day, a rising stock-market may be an almost essential condition of a satisfactory propensity to consume; and this circumstance, gener-ally overlooked until lately, obviously serves to aggravate still further the depressing effect of a decline in the marginal efficiency of capital.

When once the recovery has been started, the manner in which it feeds on itself and cumulates is obvious. But during the downward phase, when both fixed capital and stocks of materials are for the time being redundant and working-capital is being reduced, the schedule of the marginal efficiency of capital may fall so low that it can scarcely be corrected, so as to

[1] Some part of the discussion in my *Treatise on Money* [*JMK*, vol. v], Book IV, bears upon the above.

secure a satisfactory rate of new investment, by any practicable reduction in the rate of interest. Thus with markets organised and influenced as they are at present, the market estimation of the marginal efficiency of capital may suffer such enormously wide fluctuations that it cannot be sufficiently offset by corresponding fluctuations in the rate of interest. Moreover, the corresponding movements in the stock-market may, as we have seen above, depress the propensity to consume just when it is most needed. In conditions of *laissez-faire* the avoidance of wide fluctuations in employment may, therefore, prove impossible without a far-reaching change in the psychology of investment markets such as there is no reason to expect. I conclude that the duty of ordering the current volume of investment cannot safely be left in private hands.

III

The preceding analysis may appear to be in conformity with the view of those who hold that over-investment is the characteristic of the boom, that the avoidance of this over-investment is the only possible remedy for the ensuing slump, and that, whilst for the reasons given above the slump cannot be prevented by a low rate of interest, nevertheless the boom can be avoided by a high rate of interest. There is, indeed, force in the argument that a high rate of interest is much more effective against a boom than a low rate of interest against a slump.

To infer these conclusions from the above would, however, misinterpret my analysis; and would, according to my way of thinking, involve serious error. For the term over-investment is ambiguous. It may refer to investments which are destined to disappoint the expectations which prompted them or for which there is no use in conditions of severe unemployment, or it may indicate a state of affairs where every kind of capital-goods is so abundant that there is no new investment which is expected, even in conditions of full employment, to earn in the course of its life more than its replacement cost. It is only the latter state of affairs which is one of over-investment, strictly speaking, in the sense that any further investment would be a sheer waste of resources.[1] Moreover, even if over-investment in this sense was a normal characteristic of the boom, the remedy would not lie in clapping on a high rate of interest which would probably deter some useful investments and might further diminish the propensity to consume, but in taking drastic steps, by redistributing incomes or otherwise, to stimulate the propensity to consume.

According to my analysis, however, it is only in the former sense that the boom can be said to be characterised by over-investment. The situation, which I am indicating as typical, is not one in which capital is so abundant that the community as a whole has no reasonable use for any more, but where investment is being made in conditions which are unstable and cannot endure, because it is prompted by expectations which are destined to disappointment.

It may, of course, be the case—indeed it is likely to be—that the illusions of the boom cause particular types of capital-assets to be produced in such excessive abundance that some part of the output is, on any criterion, a waste of resources;—which sometimes happens, we may add, even when there is no boom. It leads, that is to say, to *misdirected* investment. But over and above this it is an essential characteristic of the boom that investments which will in fact yield, say, 2 per cent in conditions of full employment are made in the expectation of a yield of, say, 6 per cent, and are valued accordingly. When the disillusion comes, this expectation is

---

[1] On certain assumptions, however, as to the distribution of the propensity to consume through time, investment which yielded a negative return might be advantageous in the sense that, for the community as a whole, it would maximise satisfaction.

replaced by a contrary 'error of pessimism', with the result that the investments, which would in fact yield 2 per cent in conditions of full employment, are expected to yield less than nothing; and the resulting collapse of new investment then leads to a state of unemployment in which the investments, which would have yielded 2 per cent in conditions of full employment, in fact yield less than nothing. We reach a condition where there is a shortage of houses, but where nevertheless no one can afford to live in the houses that there are.

Thus the remedy for the boom is not a higher rate of interest but a lower rate of interest![1] For that may enable the so-called boom to last. The right remedy for the trade cycle is not to be found in abolishing booms and thus keeping us permanently in a semi-slump; but in abolishing slumps and thus keeping us permanently in a quasi-boom.

The boom which is destined to end in a slump is caused, therefore, by the combination of a rate of interest, which in a correct state of expectation would be too high for full employment, with a misguided state of expectation which, so long as it lasts, prevents this rate of interest from being in fact deterrent. A boom is a situation in which over-optimism triumphs over a rate of interest which, in a cooler light, would be seen to be excessive.

Except during the war, I doubt if we have any recent experience of a boom so strong that it led to full employment. In the United States employment was very satisfactory in 1928–29 on normal standards; but I have seen no evidence of a shortage of labour, except, perhaps, in the case of a few groups of highly specialised workers. Some 'bottle-necks' were reached, but output as a whole was still capable of further expansion. Nor was there over-investment in the sense that the

---

[1] See below (p. 327) for some arguments which can be urged on the other side. For, if we are precluded from making large changes in our present methods, I should agree that to raise the rate of interest during a boom may be, in conceivable circumstances, the lesser evil.

standard and equipment of housing was so high that everyone, assuming full employment, had all he wanted at a rate which would no more than cover the replacement cost, without any allowance for interest, over the life of the house; and that transport, public services and agricultural improvement had been carried to a point where further additions could not reasonably be expected to yield even their replacement cost. Quite the contrary. It would be absurd to assert of the United States in 1929 the existence of over-investment in the strict sense. The true state of affairs was of a different character. New investment during the previous five years had been, indeed, on so enormous a scale in the aggregate that the prospective yield of further additions was, coolly considered, falling rapidly. Correct foresight would have brought down the marginal efficiency of capital to an unprecedentedly low figure; so that the 'boom' could not have continued on a sound basis except with a very low long-term rate of interest, and an avoidance of misdirected investment in the particular directions which were in danger of being over-exploited. In fact, the rate of interest was high enough to deter new investment except in those particular directions which were under the influence of speculative excitement and, therefore, in special danger of being over-exploited; and a rate of interest, high enough to overcome the speculative excitement, would have checked, at the same time, every kind of reasonable new investment. Thus an increase in the rate of interest, as a remedy for the state of affairs arising out of a prolonged period of abnormally heavy new investment, belongs to the species of remedy which cures the disease by killing the patient.

It is, indeed, very possible that the prolongation of approximately full employment over a period of years would be associated in countries so wealthy as Great Britain or the United States with a volume of new investment, assuming the existing propensity to consume, so great that it would eventually lead to a state of full investment in the sense that an aggregate gross yield in excess of replacement cost could no longer be expected on a reasonable calculation from a further increment of durable goods of any type whatever. Moreover, this situation might be reached comparatively soon—say within twenty-five years or less. I must not be taken to deny this, because I assert that a state of full investment in the strict sense has never yet occurred, not even momentarily.

Furthermore, even if we were to suppose that contemporary booms are apt to be associated with a momentary condition of full investment or over-investment in the strict sense, it would still be absurd to regard a higher rate of interest as the appropriate remedy. For in this event the case of those who attribute the disease to under-consumption would be wholly established. The remedy would lie in various measures designed to increase the propensity to consume by the redistribution of incomes or otherwise; so that a given level of employment would require a smaller volume of current investment to support it.

IV

It may be convenient at this point to say a word about the important schools of thought which maintain, from various points of view, that the chronic tendency of contemporary societies to under-employment is to be traced to under-consumption;—that is to say, to social practices and to a distribution of wealth which result in a propensity to consume which is unduly low.

In existing conditions—or, at least, in the condition which existed until lately—where the volume of investment is unplanned and uncontrolled, subject to the vagaries of the marginal efficiency of capital as determined by the private judgment of individuals ignorant or speculative, and to a long-term rate of interest which seldom or never falls below a conventional level, these schools of thought are, as guides to practical policy, undoubtedly in the right. For in such conditions there is no other means of raising the average level of employment to a more satisfactory level. If it is impracticable materially to increase investment, obviously there is no means of securing a higher level of employment except by increasing consumption.

Practically I only differ from these schools of thought in thinking that they may lay a little too much emphasis on increased consumption at a time when there is still much social advantage to be obtained from increased investment. Theoretically, however, they are open to the criticism of neglecting the fact that there are *two* ways to expand output. Even if we were to decide that it would be better to increase capital more slowly and to concentrate effort on increasing consumption, we must decide this with open eyes after well considering the alternative. I am myself impressed by the great social advantages of increasing the stock of capital until it ceases to be scarce. But this is a practical judgment, not a theoretical imperative.

Moreover, I should readily concede that the wisest course is to advance on both fronts at once. Whilst aiming at a socially controlled rate of investment with a view to a progressive decline in the marginal efficiency of capital, I should support at the same time all sorts of policies for increasing the propensity to consume. For it is unlikely that full employment can be maintained, whatever we may do about investment, with the existing propensity to consume. There is room, therefore, for both policies to operate together;—to promote investment and, at the same time, to promote consumption, not merely to the level which with the existing propensity to consume would correspond to the increased investment, but to a higher level still.

If—to take round figures for the purpose of illustration—the average level of output of to-day is 15 per cent below what it would be with continuous full employment, and if 10 per cent of this output represents net investment and 90 per cent of it consumption—if, furthermore, net investment would have to rise 50 per cent in order to secure full employment with the existing propensity to consume, so that with full employment output would rise from 100 to 115, consumption from 90 to 100 and net investment from 10 to 15:—then we might aim, perhaps, at so modifying the propensity to consume that with full employment consumption would rise from 90 to 103 and net investment from 10 to 12.

V

Another school of thought finds the solution of the trade cycle, not in increasing either consumption or investment, but in diminishing the supply of labour seeking employment; i.e. by redistributing the existing volume of employment without increasing employment or output.

This seems to me to be a premature policy—much more clearly so than the plan of increasing consumption. A point comes where every individual weighs the advantages of increased leisure against increased income. But at present the evidence is, I think, strong that the great majority of individuals would prefer increased income to increased leisure; and I see no sufficient reason for compelling those who would prefer more income to enjoy more leisure.

VI

It may appear extraordinary that a school of thought should exist which finds the solution for the trade cycle in checking the boom in its early stages by a higher rate of interest. The only line of argument, along which any justification for this policy can be discovered, is that put forward by Mr D. H. Robertson, who assumes, in effect, that full employment is an impracticable ideal and that the best that we can hope for is a level of employment much more stable than at present and averaging, perhaps, a little higher.

If we rule out major changes of policy affecting either the control of investment or the propensity to consume, and assume, broadly speaking, a continuance of the existing state of affairs, it is, I think, arguable that a more advantageous average state of expectation might result from a banking policy which always nipped in the bud an incipient boom by a rate of interest high enough to deter even the most misguided optimists. The disappointment of expectation, characteristic of the slump, may lead to so much loss and waste that the average level of useful investment might be higher if a deterrent is applied. It is difficult to be sure whether or not this is correct on its own assumptions; it is a matter for practical judgment where detailed evidence is wanting. It may be that it overlooks the social advantage which accrues from the increased consumption which attends even on investment which proves to have been totally misdirected, so that even such investment may be more beneficial than no investment at all. Nevertheless, the most enlightened monetary control might find itself in difficulties, faced with a boom of the 1929 type in America, and armed with no other weapons than those possessed at that time by the Federal Reserve System; and none of the alternatives within its power might make much difference to the result. However this may be, such an outlook seems to me to be dangerously and unnecessarily defeatist. It recommends, or at least assumes, for permanent acceptance too much that is defective in our existing economic scheme.

The austere view, which would employ a high rate of interest to check at once any tendency in the level of employment to rise appreciably above the average of, say, the previous decade, is, however, more usually supported by arguments which have no foundation at all apart from confusion of mind. It flows, in some cases, from the belief that in a boom investment tends to outrun saving, and that a higher rate of interest will restore equilibrium by checking investment on the one hand and stimulating savings on the other. This implies that saving and investment can be unequal, and has, therefore, no meaning until these terms have been defined in some special sense. Or it is sometimes suggested that the increased saving which accompanies increased investment is undesirable and unjust because it is, as a rule, also associated with rising prices. But if this were so, *any* upward change in the existing level of output and employment is to be deprecated. For the rise in prices is not essentially due to the increase in investment;—it is due to the fact that in the short period supply price usually increases with increasing output, on account either of the physical fact of diminishing return or of the tendency of the cost-unit to rise in terms of money when output increases. If the conditions were those of constant supply-price, there would, of course, be no rise of prices; yet, all the same, increased saving would accompany increased investment. It is the increased output which produces the increased saving; and the rise of prices is merely a by-product of the increased output, which will occur equally if there is no increased saving but, instead, an increased propensity to consume. No one has a legitimate vested interest in being able to buy at prices which are only low because output is low.

Or, again, the evil is supposed to creep in if the increased investment has been promoted by a fall in the rate of interest engineered by an increase in the quantity of money. Yet there is no special virtue in the pre-existing rate of interest, and the new money is not 'forced' on anyone;—it is created in order to satisfy the increased liquidity-preference which corresponds to the lower rate of interest or the increased volume of transactions, and it is held by those individuals who *prefer* to hold money rather than to lend it at the lower rate of interest. Or, once more, it is suggested that a boom is characterised by 'capital consumption', which presumably means negative net investment, i.e. by an excessive propensity to consume. Unless the phenomena of the trade cycle have been confused with those of a flight from the currency such as occurred during the post-war European currency collapses, the evidence is wholly to the contrary. Moreover, even if it were so, a reduction in the rate of interest would be a more plausible remedy than a rise in the rate of interest for conditions of under-investment. I can make no sense at all of these schools of thought; except, perhaps, by supplying a tacit assumption that aggregate output is incapable of change. But a theory which assumes constant output is obviously not very serviceable for explaining the trade cycle.

VII

In the earlier studies of the trade cycle, notably by Jevons, an explanation was found in agricultural fluctuations due to the seasons, rather than in the phenomena of industry. In the light of the above theory this appears as an extremely plausible approach to the problem. For even to-day fluctuation in the stocks of agricultural products as between one year and another is one of the largest individual items amongst the causes of changes in the rate of current investment; whilst at the time when Jevons wrote—and more particularly over the period to which most of his statistics applied— this factor must have far outweighed all others.

Jevons's theory, that the trade cycle was primarily due to the fluctuations in the bounty of the harvest, can be re-stated as follows. When an exceptionally large harvest is gathered in, an important addition is usually made to the quantity carried over into later years. The proceeds of this addition are added to the current incomes of the farmers and are treated by them as income; whereas the increased carry-over involves no drain on the income-expenditure of other sections of the community but is financed out of savings. That is to say, the addition to the carry-over is an addition to current investment. This conclusion is not invalidated even if prices fall sharply. Similarly when there is a poor harvest, the carry-over is drawn upon for current consumption, so that a corresponding part of the income-expenditure of the consumers creates no current income for the farmers. That is to say, what is taken from the carry-over involves a corresponding reduction in current investment. Thus, if investment in other directions is taken to be constant, the difference in aggregate investment between a year in which there is a substantial addition to the carry-over and a year in which there is a substantial subtraction from it may be large; and in a community where agriculture is the predominant industry it will be overwhelmingly large compared with any other usual cause of investment fluctuations. Thus it is natural that we should find the upward turning-point to be marked by bountiful harvests and the downward turning-point by deficient harvests. The further theory, that there are physical causes for a regular cycle of good and bad harvests, is, of course, a different matter with which we are not concerned here.

More recently, the theory has been advanced that it is bad harvests, not good harvests, which are good for trade, either because bad harvests make the population ready to work for a smaller real reward or because the resulting redistribution of purchasing-power is held to be favourable to consumption. Needless to say, it is not these theories which I have in mind in the above description of harvest phenomena as an explanation of the trade cycle.

The agricultural causes of fluctuation are, however, much less important in the modern world for two reasons. In the first place agricultural output is a much smaller proportion of total output. And in the second place the development of a world market for most agricultural products, drawing upon both hemispheres, leads to an averaging out of the effects of good and bad seasons, the percentage fluctuation in the amount of the world harvest being far less than the percentage fluctuations in the harvests of individual countries. But in old days, when a country was mainly dependent on its own harvest, it is difficult to see any possible cause of fluctuations in investment, except war, which was in any way comparable in magnitude with changes in the carry-over of agricultural products.

Even to-day it is important to pay close attention to the part played by changes in the stocks of raw materials, both agricultural and mineral, in the determination of the rate of current investment. I should attribute the slow rate of recovery from a slump, after the turning-point has been reached, mainly to the deflationary effect of the reduction of redundant stocks to a normal level. At first the accumulation of stocks, which occurs after the boom has broken, moderates the rate of the collapse; but we have to pay for this relief later on in the damping-down of the subsequent rate of recovery. Sometimes, indeed, the reduction of stocks may have to be virtually completed before any measurable degree of recovery can be detected. For a rate of investment in other directions, which is sufficient to produce an upward movement when there is no current disinvestment in stocks to set off against it, may be quite inadequate so long as such disinvestment is still proceeding.

We have seen, I think, a signal example of this in the earlier phases of America's 'New Deal'. When President Roosevelt's substantial loan expenditure began, stocks of all kinds—and particularly of agricultural products—still stood at a very high level. The 'New Deal' partly consisted in a strenuous attempt to reduce these stocks—by curtailment of current output and in all sorts of ways. The reduction of stocks to a normal level was a necessary process—a phase which had to be endured. But so long as it lasted, namely, about two years, it constituted a substantial offset to the loan expenditure which was being incurred in other directions. Only when it had been completed was the way prepared for substantial recovery.

Recent American experience has also afforded good examples of the part played by fluctuations in the stocks of finished and unfinished goods—'inventories' as it is becoming usual to call them—in causing the minor oscillations within the main movement of the trade cycle. Manufacturers, setting industry in motion to provide for a scale of consumption which is expected to prevail some months later, are apt to make minor miscalculations, generally in the direction of running a little ahead of the facts. When they discover their mistake they have to contract for a short time to a level below that of current consumption so as to allow for the absorption of the excess inventories; and the difference of pace between running a little ahead and dropping back again has proved sufficient in its effect on the current rate of investment to display itself quite clearly against the background of the excellently complete statistics now available in the United States.

Chapter 23

NOTES ON MERCANTILISM, THE USURY LAWS, STAMPED MONEY AND THEORIES OF UNDER-CONSUMPTION

I

For some two hundred years both economic theorists and practical men did not doubt that there is a peculiar advantage to a country in a favourable balance of trade, and grave danger in an unfavourable balance, particularly if it results in an efflux of the precious metals. But for the past one hundred years there has been a remarkable divergence of opinion. The majority of statesmen and practical men in most countries, and nearly half of them even in Great Britain, the home of the opposite view, have remained faithful to the ancient doctrine; whereas almost all economic theorists have held that anxiety concerning such matters is absolutely groundless except on a very short view, since the mechanism of foreign trade is self-adjusting and attempts to interfere with it are not only futile, but greatly impoverish those who practise them because they forfeit the advantages of the international division of labour. It will be convenient, in accordance with tradition, to designate the older opinion as *mercantilism* and the newer as *free trade*, though these terms, since each of them has both a broader and a narrower signification, must be interpreted with reference to the context.

Generally speaking, modern economists have maintained not merely that there is, as a rule, a balance of gain from the international division of labour sufficient to outweigh such advantages as mercantilist practice can fairly claim, but that the mercantilist argument is based, from start to finish, on an intellectual confusion.

Marshall,[1] for example, although his references to mercantilism are not altogether unsympathetic, had no regard for their central theory as such and does not even mention those elements of truth in their contentions which I shall examine below.[2] In the same way, the theoretical concessions which free-trade economists have been ready to make in contemporary controversies, relating, for example, to the encouragement of infant industries or to the improvement of the terms of trade, are not concerned with the real substance of the mercantilist case. During the fiscal controversy of the first quarter of the present century I do not remember that any concession was ever allowed by economists to the claim that protection might increase domestic employment. It will be fairest, perhaps, to quote, as an example, what I wrote myself. So lately as 1923, as a faithful pupil of the classical school who did not at that time doubt what he had been taught and entertained on this matter no reserves at all, I wrote: 'If there is one thing that Protection can *not* do, it is to cure Unemployment... There are some arguments for Protection, based upon its securing possible but improbable advantages, to which there is no simple answer. But the claim to cure Unemployment involves the Protectionist fallacy in its grossest and crudest form.'[3] As for earlier mercantilist theory, no

---

[1] *Vide* his *Industry and Trade*, Appendix D; *Money, Credit and Commerce*, p. 130; and *Principles of Economics*, Appendix I.

[2] His view of them is well summed up in a footnote to the first edition of his *Principles*, p. 51: 'Much study has been given both in England and Germany to medieval opinions as to the relation of money to national wealth. On the whole they are to be regarded as confused through want of a clear understanding of the functions of money, rather than as wrong in consequence of a deliberate assumption that the increase in the net wealth of a nation can be effected only by an increase of the stores of the precious metals in her.'

[3] *The Nation and the Athenaeum*, November 24, 1923 [*JMK*, vol. XVIII].

intelligible account was available; and we were brought up to believe that it was little better than nonsense. So absolutely overwhelming and complete has been the domination of the classical school.

II

Let me first state in my own terms what now seems to me to be the element of scientific truth in mercantilist doctrine. We will then compare this with the actual arguments of the mercantilists. It should be understood that the advantages claimed are avowedly national advantages and are unlikely to benefit the world as a whole.

When a country is growing in wealth somewhat rapidly, the further progress of this happy state of affairs is liable to be interrupted, in conditions of *laissez-faire*, by the insufficiency of the inducements to new investment. Given the social and political environment and the national characteristics which determine the propensity to consume, the well-being of a progressive state essentially depends, for the reasons we have already explained, on the sufficiency of such inducements. They may be found either in home investment or in foreign investment (including in the latter the accumulation of the precious metals), which, between them, make up aggregate investment. In conditions in which the quantity of aggregate investment is determined by the profit motive alone, the opportunities for home investment will be governed, in the long run, by the domestic rate of interest; whilst the volume of foreign investment is necessarily determined by the size of the favourable balance of trade. Thus, in a society where there is no question of direct investment under the aegis of public authority, the economic objects, with which it is reasonable for the government to be preoccupied, are the domestic rate of interest and the balance of foreign trade.

Now, if the wage-unit is somewhat stable and not liable to spontaneous changes of significant magnitude (a condition which is almost always satisfied), if the state of liquidity-preference is somewhat stable, taken as an average of its short-period fluctuations, and if banking conventions are also stable, the rate of interest will tend to be governed by the quantity of the precious metals, measured in terms of the wage-unit, available to satisfy the community's desire for liquidity. At the same time, in an age in which substantial foreign loans and the outright ownership of wealth located abroad are scarcely practicable, increases and decreases in the quantity of the precious metals will largely depend on whether the balance of trade is favourable or unfavourable.

Thus, as it happens, a preoccupation on the part of the authorities with a favourable balance of trade served *both* purposes; and was, furthermore, the only available means of promoting them. At a time when the authorities had no direct control over the domestic rate of interest or the other inducements to home investment, measures to increase the favourable balance of trade were the only *direct* means at their disposal for increasing foreign investment; and, at the same time, the effect of a favourable balance of trade on the influx of the precious metals was their only *indirect* means of reducing the domestic rate of interest and so increasing the inducement to home investment.

There are, however, two limitations on the success of this policy which must not be overlooked. If the domestic rate of interest falls so low that the volume of investment is sufficiently stimulated to raise employment to a level which breaks through some of the critical points at which the wage-unit rises, the increase in the domestic level of costs will begin to react unfavourably on the balance of foreign trade, so that the effort to increase the latter will have overreached and defeated itself. Again, if the domestic rate of interest falls so low relatively to rates of interest elsewhere as to stimulate a volume of foreign lending which is disproportionate to the favourable balance, there may ensue an efflux of the precious metals sufficient to reverse the advantages previously obtained. The risk of one or other of these limitations becoming operative is increased in the case of a country which is large and internationally important by the fact that, in conditions where the current output of the precious metals from the mines is on a relatively small scale, an influx of money into one country means an efflux from another; so that the adverse effects of rising costs and falling rates of interest at home may be accentuated (if the mercantilist policy is pushed too far) by falling costs and rising rates of interest abroad.

The economic history of Spain in the latter part of the fifteenth and in the sixteenth centuries provides an example of a country whose foreign trade was destroyed by the effect on the wage-unit of an excessive abundance of the precious metals. Great Britain in the pre-war years of the twentieth century provides an example of a country in which the excessive facilities for foreign lending and the purchase of properties abroad frequently stood in the way of the decline in the domestic rate of interest which was required to ensure full employment at home. The history of India at all times has provided an example of a country impoverished by a preference for liquidity amounting to so strong a passion that even an enormous and chronic influx of the precious metals has been insufficient to bring down the rate of interest to a level which was compatible with the growth of real wealth.

Nevertheless, if we contemplate a society with a somewhat stable wage-unit, with national characteristics which determine the propensity to consume and the preference for liquidity, and with a monetary system which rigidly links the quantity of money to the stock of the precious metals, it will be essential for the maintenance of prosperity that the authorities should pay close attention to the state of the balance of trade. For a favourable balance, provided it is not too large, will prove extremely stimulating; whilst an unfavourable balance may soon produce a state of persistent depression.

It does not follow from this that the maximum degree of restriction of imports will promote the maximum favourable balance of trade. The earlier mercantilists laid great emphasis on this and were often to be found opposing trade restrictions because on a long view they were liable to operate adversely to a favourable balance. It is, indeed, arguable that in the special circumstances of mid-nineteenth-century Great Britain an almost complete freedom of trade was the policy most conducive to the development of a favourable balance. Contemporary experience of trade restrictions in post-war Europe offers manifold examples of ill-conceived impediments on freedom which, designed to improve the favourable balance, had in fact a contrary tendency.

For this and other reasons the reader must not reach a premature conclusion as to the *practical* policy to which our argument leads up. There are strong presumptions of a general character against trade restrictions unless they can be justified on special grounds. The advantages of the international division of labour are real and substantial, even though the classical school greatly overstressed them. The fact that the advantage which our own country gains from a favourable balance is liable to involve an equal disadvantage to some other country (a point to which the mercantilists were fully alive) means not only that great moderation is necessary, so that a country secures for itself no larger a share of the stock of the precious metals than is fair and reasonable, but also that an immoderate policy may lead to a senseless international competition for a favourable balance which injures all alike.[1] And finally, a policy of trade restrictions is a treacherous instrument even for the attainment of its ostensible object, since private interest, administrative incompetence and the intrinsic difficulty of the task may divert it into producing results directly opposite to those intended.

Thus, the weight of my criticism is directed against the inadequacy of the *theoretical* foundations of the *laissez-faire* doctrine upon which I was brought up and which for many years I taught;—against the notion that the rate of interest and the volume of investment are self-adjusting at the optimum level, so that preoccupation with the balance of trade is a waste of time. For we, the faculty of economists, prove to have been guilty of presumptuous error in treating as a puerile obsession what for centuries has been a prime object of practical statecraft.

Under the influence of this faulty theory the City of London gradually devised the most dangerous technique for the maintenance of equilibrium which can possibly be imagined, namely, the technique of bank rate coupled with a rigid parity of the foreign exchanges. For this meant that the objective of maintaining a domestic rate of interest consistent with full employment was wholly ruled out. Since, in practice, it is impossible to neglect the balance of payments, a means of controlling it was evolved which, instead of protecting the domestic rate of interest, sacrificed it to the operation of blind forces. Recently, practical bankers in London have learnt much, and one can almost hope that in Great Britain the technique of bank rate will never be used again to protect the foreign balance in conditions in which it is likely to cause unemployment at home.

Regarded as the theory of the individual firm and

---

[1] The remedy of an elastic wage-unit, so that a depression is met by a reduction of wages, is liable, for the same reason, to be a means of benefiting ourselves at the expense of our neighbours.

of the distribution of the product resulting from the employment of a given quantity of resources, the classical theory has made a contribution to economic thinking which cannot be impugned. It is impossible to think clearly on the subject without this theory as a part of one's apparatus of thought. I must not be supposed to question this in calling attention to their neglect of what was valuable in their predecessors. Nevertheless, as a contribution to statecraft, which is concerned with the economic system as a whole and with securing the optimum employment of the system's entire resources, the methods of the early pioneers of economic thinking in the sixteenth and seventeenth centuries may have attained to fragments of practical wisdom which the unrealistic abstractions of Ricardo first forgot and then obliterated. There was wisdom in their intense preoccupation with keeping down the rate of interest by means of usury laws (to which we will return later in this chapter), by maintaining the domestic stock of money and by discouraging rises in the wage-unit; and in their readiness in the last resort to restore the stock of money by devaluation, if it had become plainly deficient through an unavoidable foreign drain, a rise in the wage-unit,[1] or any other cause.

III

The early pioneers of economic thinking may have hit upon their maxims of practical wisdom without having had much cognisance of the underlying theoretical grounds. Let us, therefore, examine briefly the reasons they gave as well as what they recommended. This is

[1] Experience since the age of Solon at least, and probably, if we had the statistics, for many centuries before that, indicates what a knowledge of human nature would lead us to expect, namely, that there is a steady tendency for the wage-unit to rise over long periods of time and that it can be reduced only amidst the decay and dissolution of economic society. Thus, apart altogether from progress and increasing population, a gradually increasing stock of money has proved imperative.

made easy by reference to Professor Heckscher's great work on *Mercantilism*, in which the essential characteristics of economic thought over a period of two centuries are made available for the first time to the general economic reader. The quotations which follow are mainly taken from his pages.[1]

(1) Mercantilists' thought never supposed that there was a self-adjusting tendency by which the rate of interest would be established at the appropriate level. On the contrary they were emphatic that an unduly high rate of interest was the main obstacle to the growth of wealth; and they were even aware that the rate of interest depended on liquidity-preference and the quantity of money. They were concerned both with diminishing liquidity-preference and with increasing the quantity of money, and several of them made it clear that their preoccupation with increasing the quantity of money was due to their desire to diminish the rate of interest. Professor Heckscher sums up this aspect of their theory as follows:

> The position of the more perspicacious mercantilists was in this respect, as in many others, perfectly clear within certain limits. For them, money was—to use the terminology of to-day—a factor of production, on the same footing as land, sometimes regarded as 'artificial' wealth as distinct from the 'natural' wealth; interest on capital was the payment for the renting of money similar to rent for land. In so far as mercantilists sought to discover objective reasons for the height of the rate of interest—and they did so more and more during this period—they found such reasons in the total quantity of money. From the abundant material available, only the most typical examples will be selected, so as to demonstrate first and foremost how lasting this notion was, how deep-rooted and independent of practical considerations.
>
> Both of the protagonists in the struggle over monetary

[1] They are the more suitable for my purpose because Prof. Heckscher is himself an adherent, on the whole, of the classical theory and much less sympathetic to the mercantilist theories than I am. Thus there is no risk that his choice of quotations has been biassed in any way by a desire to illustrate their wisdom.

policy and the East India trade in the early 1620's in England were in entire agreement on this point. Gerard Malynes stated, giving detailed reason for his assertion, that 'Plenty of money decreaseth usury in price or rate' (*Lex Mercatoria* and *Maintenance of Free Trade*, 1622). His truculent and rather unscrupulous adversary, Edward Misselden, replied that 'The remedy for Usury may be plenty of money' (*Free Trade or the Meanes to make Trade Florish*, same year). Of the leading writers of half a century later, Child, the omnipotent leader of the East India Company and its most skilful advocate, discussed (1668) the question of how far the legal maximum rate of interest, which he emphatically demanded, would result in drawing 'the money' of the Dutch away from England. He found a remedy for this dreaded disadvantage in the easier transference of bills of debt, if these were used as currency, for this, he said, 'will certainly supply the defect of at least one-half of all the ready money we have in use in the nation'. Petty, the other writer, who was entirely unaffected by the clash of interests, was in agreement with the rest when he explained the 'natural' fall in the rate of interest from 10 per cent to 6 per cent by the increase in the amount of money (*Political Arithmetick*, 1676), and advised lending at interest as an appropriate remedy for a country with too much 'Coin' (*Quantulumcunque concerning Money*, 1682).

This reasoning, naturally enough, was by no means confined to England. Several years later (1701 and 1706), for example, French merchants and statesmen complained of the prevailing scarcity of coin (*disette des espèces*) as the cause of the high interest rates, and they were anxious to lower the rate of usury by increasing the circulation of money.[1]

The great Locke was, perhaps, the first to express in abstract terms the relationship between the rate of interest and the quantity of money in his controversy with Petty.[2] He was opposing Petty's proposal of a maximum rate of interest on the ground that it was as impracticable as to fix a maximum rent for land, since 'the natural Value of Money, as it is apt to yield such

[1] Heckscher, *Mercantilism*, vol. ii. pp. 200, 201, very slightly abridged.
[2] *Some Considerations of the Consequences of the Lowering of Interest and Raising the Value of Money*, 1692, but written some years previously.

an yearly Income by Interest, depends on the whole quantity of the then passing Money of the Kingdom, in proportion to the whole Trade of the Kingdom (i.e. the general Vent of all the commodities)'.[1] Locke explains that money has two values: (1) its value in use which is given by the rate of interest 'and in this it has the Nature of Land, the Income of one being called Rent, of the other, Use[2]', and (2) its value in exchange 'and in this it has the Nature of a Commodity', its value in exchange 'depending only on the Plenty or Scarcity of Money in proportion to the Plenty or Scarcity of those things and not on what Interest shall be'. Thus Locke was the parent of twin quantity theories. In the first place he held that the rate of interest depended on the proportion of the quantity of money (allowing for the velocity of circulation) to the total value of trade. In the second place he held that the value of money in exchange depended on the proportion of the quantity of money to the total volume of goods in the market. But—standing with one foot in the mercantilist world and with one foot in the classical world[3]—he was confused concerning the relation between these two proportions, and he overlooked altogether the possibility of *fluctuations* in liquidity-preference. He was, however, eager to explain that a

---

[1]   He adds: 'not barely on the quantity of money but the quickness of its circulation'.

[2]   'Use' being, of course, old-fashioned English for 'interest'.

[3]   Hume a little later had a foot and a half in the classical world. For Hume began the practice amongst economists of stressing the importance of the equilibrium position as compared with the ever-shifting transition towards it, though he was still enough of a mercantilist not to overlook the fact that it is in the transition that we actually have our being: 'It is only in this interval or intermediate situation, between the acquisition of money and a rise of prices, that the increasing quantity of gold and silver is favourable to industry..It is of no manner of consequence, with regard to the domestic happiness of a state, whether money be in a greater or less quantity. The good policy of the magistrate consists only in keeping it, if possible, still increasing; because by that means he keeps alive a spirit of industry in the nation, and increases the state of labour in which consists all real power and riches. A nation, whose money decreases, is actually, at that time, weaker and more miserable than another nation, which possesses no more money but is on the increasing trend.' (Essay *On Money*, 1752.)

reduction in the rate of interest has no *direct* effect on the price-level and affects prices 'only as the Change of Interest in Trade conduces to the bringing in or carrying out Money or Commodity, and so in time varying their Proportion here in England from what it was before', i.e. if the reduction in the rate of interest leads to the export of cash or an increase in output. But he never, I think, proceeds to a genuine synthesis.[1]

How easily the mercantilist mind distinguished between the rate of interest and the marginal efficiency of capital is illustrated by a passage (printed in 1621) which Locke quotes from *A Letter to a Friend concerning Usury*: 'High Interest decays Trade. The advantage from Interest is greater than the Profit from Trade, which makes the rich Merchants give over, and put out their Stock to Interest, and the lesser Merchants Break.' Fortrey (*England's Interest and Improvement*, 1663) affords another example of the stress laid on a low rate of interest as a means of increasing wealth.

The mercantilists did not overlook the point that, if an excessive liquidity-preference were to withdraw the influx of precious metals into hoards, the advantage to the rate of interest would be lost. In some cases (e.g. Mun) the object of enhancing the power of the State led them, nevertheless, to advocate the accumulation of state treasure. But others frankly opposed this policy:

> Schrötter, for instance, employed the usual mercantilist arguments in drawing a lurid picture of how the circulation in the country would be robbed of all its money through a greatly increasing state treasury...he, too, drew a perfectly logical parallel between the accumulation of treasure by the

[1] It illustrates the completeness with which the mercantilist view, that interest *means* interest on money (the view which is, as it now seems to me, indubitably correct), has dropt out, that Prof. Heckscher, as a good classical economist, sums up his account of Locke's theory with the comment—'Locke's argument would be irrefutable...if interest really were synonymous with the price for the loan of money; as this is not so, it is entirely irrelevant' (*op. cit.* vol. ii. p. 204).

monasteries and the export surplus of precious metals, which, to him, was indeed the worst possible thing which he could think of. Davenant explained the extreme poverty of many Eastern nations—who were believed to have more gold and silver than any other countries in the world—by the fact that treasure 'is suffered to stagnate in the Princes' Coffers'... If hoarding by the state was considered, at best, a doubtful boon, and often a great danger, it goes without saying that private hoarding was to be shunned like the pest. It was one of the tendencies against which innumerable mercantilist writers thundered, and I do not think it would be possible to find a single dissentient voice.[1]

(2) The mercantilists were aware of the fallacy of cheapness and the danger that excessive competition may turn the terms of trade against a country. Thus Malynes wrote in his *Lex Mercatoria* (1622): 'Strive not to undersell others to the hurt of the Commonwealth, under colour to increase trade: for trade doth not increase when commodities are good cheap, because the cheapness proceedeth of the small request and scarcity of money, which maketh things cheap: so that the contrary augmenteth trade when there is plenty of money, and commodities become dearer being in request'.[2] Professor Heckscher sums up as follows this strand in mercantilist thought:

> In the course of a century and a half this standpoint was formulated again and again in this way, that a country with relatively less money than other countries must 'sell cheap and buy dear'...
> Even in the original edition of the *Discourse of the Common Weal*, that is in the middle of the 16th century, this attitude was already manifested. Hales said, in fact, 'And yet if strangers should be content to take but our wares for theirs, what should let them to advance the price of other things (meaning: among others, such as we buy from them), though ours were good cheap unto them? And then shall we be still losers, and they at the winning hand with us, while they sell dear and yet buy ours good cheap, and consequently enrich

---

[1] Heckscher, *op. cit.* vol. ii. pp. 210, 211.
[2] Heckscher, *op. cit.* vol. ii. p. 228.

themselves and impoverish us. Yet had I rather advance our wares in price, as they advance theirs, as we now do; though some be losers thereby, and yet not so many as should be the other way.' On this point he had the unqualified approval of his editor several decades later (1581). In the 17th century, this attitude recurred again without any fundamental change in significance. Thus, Malynes believed this unfortunate position to be the result of what he dreaded above all things, *i.e.* a foreign under-valuation of the English exchange...The same conception then recurred continually. In his *Verbum Sapienti* (written 1665, published 1691), Petty believed that the violent efforts to increase the quantity of money could only cease 'when we have certainly more money than any of our Neighbour States (though never so little), both in Arithmetical and Geometrical proportion'. During the period between the writing and the publication of this work, Coke declared, 'If our Treasure were more than our Neighbouring Nations, I did not care whether we had one fifth part of the Treasure we now have' (1675).[1]

(3) The mercantilists were the originals of 'the fear of goods' and the scarcity of money as causes of unemployment which the classicals were to denounce two centuries later as an absurdity:

> One of the earliest instances of the application of the unemployment argument as a reason for the prohibition of imports is to be found in Florence in the year 1426....The English legislation on the matter goes back to at least 1455.... An almost contemporary French decree of 1466, forming the basis of the silk industry of Lyons, later to become so famous, was less interesting in so far as it was not actually directed against foreign goods. But it, too, mentioned the possibility of giving work to tens of thousands of unemployed men and women. It is seen how very much this argument was in the air at the time...
>
> The first great discussion of this matter, as of nearly all social and economic problems, occurred in England in the middle of the 16th century or rather earlier, during the reigns of Henry VIII and Edward VI. In this connection we cannot but mention a series of writings, written apparently at the latest in the 1530's, two of which at any rate are believed

[1] Heckscher, *op. cit.* vol. ii. p. 235.

to have been by Clement Armstrong...He formulates it, for example, in the following terms: 'By reason of great abundance of strange merchandises and wares brought yearly into England hath not only caused scarcity of money, but hath destroyed all handicrafts, whereby great number of common people should have works to get money to pay for their meat and drink, which of very necessity must live idly and beg and steal'.[1]

The best instance to my knowledge of a typically mercantilist discussion of a state of affairs of this kind is the debates in the English House of Commons concerning the scarcity of money, which occurred in 1621, when a serious depression had set in, particularly in the cloth export. The conditions were described very clearly by one of the most influential members of parliament, Sir Edwin Sandys. He stated that the farmer and the artificer had to suffer almost everywhere, that looms were standing idle for want of money in the country, and that peasants were forced to repudiate their contracts, 'not (thanks be to God) for want of fruits of the earth, but for want of money'. The situation led to detailed enquiries into where the money could have got to, the want of which was felt so bitterly. Numerous attacks were directed against all persons who were supposed to have contributed either to an export (export surplus) of precious metals, or to their disappearance on account of corresponding activities within the country.[2]

Mercantilists were conscious that their policy, as Professor Heckscher puts it, 'killed two birds with one stone'. 'On the one hand the country was rid of an unwelcome surplus of goods, which was believed to result in unemployment, while on the other the total stock of money in the country was increased',[3] with the resulting advantages of a fall in the rate of interest.

It is impossible to study the notions to which the mercantilists were led by their actual experiences, without perceiving that there has been a chronic tendency throughout human history for the propensity to save to be stronger than the inducement to invest. The

[1] Heckscher, *op. cit.* vol. ii. p. 122.
[2] Heckscher, *op. cit.* vol. ii. p. 223.
[3] Heckscher, *op. cit.* vol. ii. p. 178.

weakness of the inducement to invest has been at all times the key to the economic problem. To-day the explanation of the weakness of this inducement may chiefly lie in the extent of existing accumulations; whereas, formerly, risks and hazards of all kinds may have played a larger part. But the result is the same. The desire of the individual to augment his personal wealth by abstaining from consumption has usually been stronger than the inducement to the entrepreneur to augment the national wealth by employing labour on the construction of durable assets.

(4) The mercantilists were under no illusions as to the nationalistic character of their policies and their tendency to promote war. It was *national* advantage and *relative* strength at which they were admittedly aiming.[1]

We may criticise them for the apparent indifference with which they accepted this inevitable consequence of an international monetary system. But intellectually their realism is much preferable to the confused thinking of contemporary advocates of an international fixed gold standard and *laissez-faire* in international lending, who believe that it is precisely these policies which will best promote peace.

For in an economy subject to money contracts and customs more or less fixed over an appreciable period of time, where the quantity of the domestic circulation and the domestic rate of interest are primarily determined by the balance of payments, as they were in Great Britain before the war, there is no orthodox means open to the authorities for countering unemployment at home except by struggling for an export surplus and

---

[1] '*Within* the stage, mercantilism pursued thoroughgoing dynamic ends. But the important thing is that this was bound up with a static conception of the total economic resources in the world; for this it was that created that fundamental disharmony which sustained the endless commercial wars.... This was the tragedy of mercantilism. Both the Middle Ages with their universal static ideal and *laissez-faire* with its universal dynamic ideal avoided this consequence' (Heckscher, *op. cit.* vol. ii. pp. 25, 26).

an import of the monetary metal at the expense of their neighbours. Never in history was there a method devised of such efficacy for setting each country's advantage at variance with its neighbours' as the international gold (or, formerly, silver) standard. For it made domestic prosperity directly dependent on a competitive pursuit of markets and a competitive appetite for the precious metals. When by happy accident the new supplies of gold and silver were comparatively abundant, the struggle might be somewhat abated. But with the growth of wealth and the diminishing marginal propensity to consume, it has tended to become increasingly internecine. The part played by orthodox economists, whose common sense has been insufficient to check their faulty logic, has been disastrous to the latest act. For when in their blind struggle for an escape, some countries have thrown off the obligations which had previously rendered impossible an autonomous rate of interest, these economists have taught that a restoration of the former shackles is a necessary first step to a general recovery.

In truth the opposite holds good. It is the policy of an autonomous rate of interest, unimpeded by international preoccupations, and of a national investment programme directed to an optimum level of domestic employment which is twice blessed in the sense that it helps ourselves and our neighbours at the same time. And it is the simultaneous pursuit of these policies by all countries together which is capable of restoring economic health and strength internationally, whether we measure it by the level of domestic employment or by the volume of international trade.[1]

---

[1] The consistent appreciation of this truth by the International Labour Office, first under Albert Thomas and subsequently under Mr H. B. Butler, has stood out conspicuously amongst the pronouncements of the numerous post-war international bodies.

I V

The mercantilists perceived the existence of the problem without being able to push their analysis to the point of solving it. But the classical school ignored the problem, as a consequence of introducing into their premisses conditions which involved its non-existence; with the result of creating a cleavage between the conclusions of economic theory and those of common sense. The extraordinary achievement of the classical theory was to overcome the beliefs of the 'natural man' and, at the same time, to be wrong. As Professor Heckscher expresses it:

> If, then, the underlying attitude towards money and the material from which money was created did not alter in the period between the Crusades and the 18th century, it follows that we are dealing with deep-rooted notions. Perhaps the same notions have persisted even beyond the 500 years included in that period, even though not nearly to the same degree as the 'fear of goods'...With the exception of the period of *laissez-faire*, no age has been free from these ideas. It was only the unique intellectual tenacity of *laissez-faire* that for a time overcame the beliefs of the 'natural man' on this point.[1]
>
> It required the unqualified faith of doctrinaire *laissez-faire* to wipe out the 'fear of goods'...[which] is the most natural attitude of the 'natural man' in a money economy. Free Trade denied the existence of factors which appeared to be obvious, and was doomed to be discredited in the eyes of the man in the street as soon as *laissez-faire* could no longer hold the minds of men enchained in its ideology.[2]

I remember Bonar Law's mingled rage and perplexity in face of the economists, because they were denying what was obvious. He was deeply troubled for an explanation. One recurs to the analogy between

---

[1] Heckscher, *op. cit.* vol. ii. pp. 176-7.
[2] *Op. cit.* vol. ii. p. 335.

the sway of the classical school of economic theory and that of certain religions. For it is a far greater exercise of the potency of an idea to exorcise the obvious than to introduce into men's common notions the recondite and the remote.

V

There remains an allied, but distinct, matter where for centuries, indeed for several millenniums, enlightened opinion held for certain and obvious a doctrine which the classical school has repudiated as childish, but which deserves rehabilitation and honour. I mean the doctrine that the rate of interest is not self-adjusting at a level best suited to the social advantage but constantly tends to rise too high, so that a wise government is concerned to curb it by statute and custom and even by invoking the sanctions of the moral law.

Provisions against usury are amongst the most ancient economic practices of which we have record. The destruction of the inducement to invest by an excessive liquidity-preference was the outstanding evil, the prime impediment to the growth of wealth, in the ancient and medieval worlds. And naturally so, since certain of the risks and hazards of economic life diminish the marginal efficiency of capital whilst others serve to increase the preference for liquidity. In a world, therefore, which no one reckoned to be safe, it was almost inevitable that the rate of interest, unless it was curbed by every instrument at the disposal of society, would rise too high to permit of an adequate inducement to invest.

I was brought up to believe that the attitude of the Medieval Church to the rate of interest was inherently absurd, and that the subtle discussions aimed at distinguishing the return on money-loans from the return to active investment were merely jesuitical attempts to find a practical escape from a foolish theory. But I now read these discussions as an honest intellectual effort to keep separate what the classical theory has inextricably confused together, namely, the rate of interest and the marginal efficiency of capital. For it now seems clear that the disquisitions of the schoolmen were directed towards the elucidation of a formula which should allow the schedule of the marginal efficiency of capital to be high, whilst using rule and custom and the moral law to keep down the rate of interest.

Even Adam Smith was extremely moderate in his attitude to the usury laws. For he was well aware that individual savings may be absorbed either by investment or by debts, and that there is no security that they will find an outlet in the former. Furthermore, he favoured a low rate of interest as increasing the chance of savings finding their outlet in new investment rather than in debts; and for this reason, in a passage for which he was severely taken to task by Bentham,[1] he defended a moderate application of the usury laws.[2] Moreover, Bentham's criticisms were mainly on the ground that Adam Smith's Scotch caution was too severe on 'projectors' and that a maximum rate of interest would leave too little margin for the reward of legitimate and socially advisable risks. For Bentham understood by *projectors* 'all such persons, as, in the pursuit of wealth, or even of any other object, endeavour, by the assistance of wealth, to strike into any channel of invention...upon all such persons as, in the line of any of their pursuits, aim at anything that can be called *improvement*...It falls, in short, upon every application of the human powers, in which ingenuity stands in need of wealth for its assistance.' Of course Bentham is right in protesting against laws which stand in the way of taking legitimate risks. 'A prudent man', Bentham continues, 'will not, in these circumstances, pick out the good projects

[1] In his *Letter to Adam Smith* appended to his *Defence of Usury.*
[2] *Wealth of Nations*, Book II, chap. 4.

from the bad, for he will not meddle with projects at all.'[1]

It may be doubted, perhaps, whether the above is just what Adam Smith intended by his term. Or is it that we are hearing in Bentham (though writing in March 1787 from 'Crichoff in White Russia') the voice of nineteenth-century England speaking to the eighteenth? For nothing short of the exuberance of the greatest age of the inducement to investment could have made it possible to lose sight of the theoretical possibility of its insufficiency.

VI

It is convenient to mention at this point the strange, unduly neglected prophet Silvio Gesell (1862-1930), whose work contains flashes of deep insight and who only just failed to reach down to the essence of the matter. In the post-war years his devotees bombarded me with copies of his works; yet, owing to certain palpable defects in the argument, I entirely failed to discover their merit. As is often the case with imperfectly analysed intuitions, their significance only became apparent after I had reached my own conclusions in my own way. Meanwhile, like other academic economists, I treated his profoundly original strivings as being no better than those of a crank. Since few of the readers of this book are likely to be well acquainted with the significance of Gesell, I will give to him what would be otherwise a disproportionate space.

Gesell was a successful German[2] merchant in

[1] Having started to quote Bentham in this context, I must remind the reader of his finest passage: 'The career of art, the great road which receives the footsteps of projectors, may be considered as a vast, and perhaps unbounded, plain, bestrewed with gulphs, such as Curtius was swallowed up in. Each requires a human victim to fall into it ere it can close, but when it once closes, it closes to open no more, and so much of the path is safe to those who follow.'

[2] Born near the Luxembourg frontier of a German father and a French mother.

Buenos Aires who was led to the study of monetary problems by the crisis of the late 'eighties, which was especially violent in the Argentine, his first work, *Die Reformation im Münzwesen als Brücke zum socialen Staat*, being published in Buenos Aires in 1891. His fundamental ideas on money were published in Buenos Aires in the same year under the title *Nervus rerum*, and many books and pamphlets followed until he retired to Switzerland in 1906 as a man of some means, able to devote the last decades of his life to the two most delightful occupations open to those who do not have to earn their living, authorship and experimental farming.

The first section of his standard work was published in 1906 at Les Hauts Geneveys, Switzerland, under the title *Die Verwirklichung des Rechtes auf dem vollen Arbeitsertrag*, and the second section in 1911 at Berlin under the title *Die neue Lehre vom Zins*. The two together were published in Berlin and in Switzerland during the war (1916) and reached a sixth edition during his lifetime under the title *Die natürliche Wirtschaftsordnung durch Freiland und Freigeld*, the English version (translated by Mr Philip Pye) being called *The Natural Economic Order*. In April 1919 Gesell joined the short-lived Soviet cabinet of Bavaria as their Minister of Finance, being subsequently tried by court-martial. The last decade of his life was spent in Berlin and Switzerland and devoted to propaganda. Gesell, drawing to himself the semi-religious fervour which had formerly centred round Henry George, became the revered prophet of a cult with many thousand disciples throughout the world. The first international convention of the Swiss and German Freiland-Freigeld Bund and similar organisations from many countries was held in Basle in 1923. Since his death in 1930 much of the peculiar type of fervour which doctrines such as his are capable of exciting has been diverted to other (in my opinion less eminent) prophets. Dr Büchi is the leader of the movement in England, but its literature seems to be distributed from San Antonio, Texas, its main strength lying to-day in the United States, where Professor Irving Fisher, alone amongst academic economists, has recognised its significance.

In spite of the prophetic trappings with which his devotees have decorated him, Gesell's main book is written in cool, scientific language; though it is suffused throughout by a more passionate, a more emotional devotion to social justice than some think decent in a scientist. The part which derives from Henry George,[1] though doubtless an important source of the movement's strength, is of altogether secondary interest. The purpose of the book as a whole may be described as the establishment of an anti-Marxian socialism, a reaction against *laissez-faire* built on theoretical foundations totally unlike those of Marx in being based on a repudiation instead of on an acceptance of the classical hypotheses, and on an unfettering of competition instead of its abolition. I believe that the future will learn more from the spirit of Gesell than from that of Marx. The preface to *The Natural Economic Order* will indicate to the reader, if he will refer to it, the moral quality of Gesell. The answer to Marxism is, I think, to be found along the lines of this preface.

Gesell's specific contribution to the theory of money and interest is as follows. In the first place, he distinguishes clearly between the rate of interest and the marginal efficiency of capital, and he argues that it is the rate of interest which sets a limit to the rate of growth of real capital. Next, he points out that the rate of interest is a purely monetary phenomenon and that the peculiarity of money, from which flows the significance of the money rate of interest, lies in the fact that its ownership as a means of storing wealth involves the holder in negligible carrying charges, and that forms of wealth, such as stocks of commodities

---

[1] Gesell differed from George in recommending the payment of compensation when the land is nationalised.

355

THE GENERAL THEORY OF EMPLOYMENT

which do involve carrying charges, in fact yield a return because of the standard set by money. He cites the comparative stability of the rate of interest throughout the ages as evidence that it cannot depend on purely physical characters, inasmuch as the variation of the latter from one epoch to another must have been incalculably greater than the observed changes in the rate of interest; i.e. (in my terminology) the rate of interest, which depends on constant psychological characters, has remained stable, whilst the widely fluctuating characters, which primarily determine the schedule of the marginal efficiency of capital, have determined not the rate of interest but the rate at which the (more or less) given rate of interest allows the stock of real capital to grow.

But there is a great defect in Gesell's theory. He shows how it is only the existence of a rate of money interest which allows a yield to be obtained from lending out stocks of commodities. His dialogue between Robinson Crusoe and a stranger[1] is a most excellent economic parable—as good as anything of the kind that has been written—to demonstrate this point. But, having given the reason why the money-rate of interest unlike most commodity rates of interest cannot be negative, he altogether overlooks the need of an explanation why the money-rate of interest is positive, and he fails to explain why the money-rate of interest is not governed (as the classical school maintains) by the standard set by the yield on productive capital. This is because the notion of liquidity-preference had escaped him. He has constructed only half a theory of the rate of interest.

The incompleteness of his theory is doubtless the explanation of his work having suffered neglect at the hands of the academic world. Nevertheless he had carried his theory far enough to lead him to a practical recommendation, which may carry with it the essence

---

[1] *The Natural Economic Order*, pp. 297 et seq.

of what is needed, though it is not feasible in the form in which he proposed it. He argues that the growth of real capital is held back by the money-rate of interest, and that if this brake were removed the growth of real capital would be, in the modern world, so rapid that a zero money-rate of interest would probably be justified, not indeed forthwith, but within a comparatively short period of time. Thus the prime necessity is to reduce the money-rate of interest, and this, he pointed out, can be effected by causing money to incur carrying-costs just like other stocks of barren goods. This led him to the famous prescription of 'stamped' money, with which his name is chiefly associated and which has received the blessing of Professor Irving Fisher. According to this proposal currency notes (though it would clearly need to apply as well to some forms at least of bank-money) would only retain their value by being stamped each month, like an insurance card, with stamps purchased at a post office. The cost of the stamps could, of course, be fixed at any appropriate figure. According to my theory it should be roughly equal to the excess of the money-rate of interest (apart from the stamps) over the marginal efficiency of capital corresponding to a rate of new investment compatible with full employment. The actual charge suggested by Gesell was 1 per mil. per week, equivalent to 5·2 per cent per annum. This would be too high in existing conditions, but the correct figure, which would have to be changed from time to time, could only be reached by trial and error.

The idea behind stamped money is sound. It is, indeed, possible that means might be found to apply it in practice on a modest scale. But there are many difficulties which Gesell did not face. In particular, he was unaware that money was not unique in having a liquidity-premium attached to it, but differed only in degree from many other articles, deriving its importance from having a *greater* liquidity-premium than any other article. Thus if currency notes were to be deprived of their liquidity-premium by the stamping system, a long series of substitutes would step into their shoes—bank-money, debts at call, foreign money, jewellery and the precious metals generally, and so forth. As I have mentioned above, there have been times when it was probably the craving for the ownership of land, independently of its yield, which served to keep up the rate of interest;—though under Gesell's system this possibility would have been eliminated by land nationalisation.

The theories which we have examined above are directed, in substance, to the constituent of effective demand which depends on the sufficiency of the inducement to invest. It is no new thing, however, to ascribe the evils of unemployment to the insufficiency of the other constituent, namely, the insufficiency of the propensity to consume. But this alternative explanation of the economic evils of the day—equally unpopular with the classical economists—played a much smaller part in sixteenth- and seventeenth-century thinking and has only gathered force in comparatively recent times.

Though complaints of under-consumption were a very subsidiary aspect of mercantilist thought, Professor Heckscher quotes a number of examples of what he calls 'the deep-rooted belief in the utility of luxury and the evil of thrift. Thrift, in fact, was regarded as the cause of unemployment, and for two reasons: in the first place, because real income was believed to diminish by the amount of money which did not enter into exchange, and secondly, because saving was believed to withdraw money from circulation.'[1] In 1598 Laffemas (*Les Trésors et richesses pour mettre l'Estat en Splendeur*) denounced the objectors to the use of

---

[1] Heckscher, *op. cit.* vol. ii. p. 208.

French silks on the ground that all purchasers of French luxury goods created a livelihood for the poor, whereas the miser caused them to die in distress.[1] In 1662 Petty justified 'entertainments, magnificent shews, triumphal arches, etc.', on the ground that their costs flowed back into the pockets of brewers, bakers, tailors, shoemakers and so forth. Fortrey justified 'excess of apparel'. Von Schrötter (1686) deprecated sumptuary regulations and declared that he would wish that display in clothing and the like were even greater. Barbon (1690) wrote that 'Prodigality is a vice that is prejudicial to the Man, but not to trade...Covetousness is a Vice, prejudicial both to Man and Trade.'[2] In 1695 Cary argued that if everybody spent more, all would obtain larger incomes 'and might then live more plentifully'.[3]

But it was by Bernard Mandeville's *Fable of the Bees* that Barbon's opinion was mainly popularised, a book convicted as a nuisance by the grand jury of Middlesex in 1723, which stands out in the history of the moral sciences for its scandalous reputation. Only one man is recorded as having spoken a good word for it, namely Dr Johnson, who declared that it did not puzzle him, but 'opened his eyes into real life very much'. The nature of the book's wickedness can be best conveyed by Leslie Stephen's summary in the *Dictionary of National Biography*:

> Mandeville gave great offence by this book, in which a cynical system of morality was made attractive by ingenious paradoxes...His doctrine that prosperity was increased by expenditure rather than by saving fell in with many current economic fallacies not yet extinct.[4] Assuming with the

---

[1] *Op. cit.* vol. ii. p. 290. 　　　[2] *Op. cit.* vol. ii. p. 291.

[3] *Op. cit.* vol. ii. p. 209.

[4] In his *History of English Thought in the Eighteenth Century* Stephen wrote (p. 297) in speaking of 'the fallacy made celebrated by Mandeville' that 'the complete confutation of it lies in the doctrine—so rarely understood that its complete apprehension is, perhaps, the best test of an economist—that demand for commodities is not demand for labour'.

ascetics that human desires were essentially evil and therefore produced 'private vices' and assuming with the common view that wealth was a 'public benefit', he easily showed that all civilisation implied the development of vicious propensities...

The text of the *Fable of the Bees* is an allegorical poem—'The Grumbling Hive, or Knaves turned honest', in which is set forth the appalling plight of a prosperous community in which all the citizens suddenly take it into their heads to abandon luxurious living, and the State to cut down armaments, in the interests of Saving:

> No Honour now could be content,
> To live and owe for what was spent,
> Liv'ries in Broker's shops are hung;
> They part with Coaches for a song;
> Sell stately Horses by whole sets;
> and Country-Houses to pay debts.
> Vain cost is shunn'd as moral Fraud;
> They have no Forces kept Abroad;
> Laugh at th' Esteem of Foreigners,
> And empty Glory got by Wars;
> They fight, but for their Country's sake,
> When Right or Liberty's at Stake.

The haughty Chloe

> Contracts th' expensive Bill of Fare,
> And wears her strong Suit a whole Year.

And what is the result?—

> Now mind the glorious Hive, and see
> How Honesty and Trade agree:
> The Shew is gone, it thins apace;
> And looks with quite another Face,
> For 'twas not only they that went,
> By whom vast sums were yearly spent;
> But Multitudes that lived on them,
> Were daily forc'd to do the same.
> In vain to other Trades they'd fly;
> All were o'er-stocked accordingly.
> The price of Land and Houses falls;
> Mirac'lous Palaces whose Walls,

> Like those of Thebes, were rais'd by Play,
> Are to be let...
> The Building Trade is quite destroy'd,
> Artificers are not employ'd;
> No limner for his Art is fam'd,
> Stone-cutters, Carvers are not nam'd.

So 'The Moral' is:

> Bare Virtue can't make Nations live
> In Splendour. They that would revive
> A Golden Age, must be as free,
> For Acorns as for Honesty.

Two extracts from the commentary which follows the allegory will show that the above was not without a theoretical basis:

> As this prudent economy, which some people call *Saving*, is in private families the most certain method to increase an estate, so some imagine that, whether a country be barren or fruitful, the same method if generally pursued (which they think practicable) will have the same effect upon a whole nation, and that, for example, the English might be much richer than they are, if they would be as frugal as some of their neighbours. This, I think, is an error.[1]

On the contrary, Mandeville concludes:

> The great art to make a nation happy, and what we call flourishing, consists in giving everybody an opportunity of being employed; which to compass, let a Government's first care be to promote as great a variety of Manufactures, Arts and Handicrafts as human wit can invent; and the second to encourage Agriculture and Fishery in all their branches, that the whole Earth may be forced to exert itself as well as Man. It is from this Policy and not from the trifling regulations of Lavishness and Frugality that the greatness and felicity of Nations must be expected; for let the value of Gold and Silver rise or fall, the enjoyment of all Societies will ever depend upon the Fruits of the Earth and the Labour of the People; both which joined together are a more certain, a more inexhaustible

[1] Compare Adam Smith, the forerunner of the classical school, who wrote, 'What is prudence in the conduct of every private family can scarce be folly in that of a great Kingdom'—probably with reference to the above passage from Mandeville.

and a more real Treasure than the Gold of Brazil or the Silver of Potosi.

No wonder that such wicked sentiments called down the opprobrium of two centuries of moralists and economists who felt much more virtuous in possession of their austere doctrine that no sound remedy was discoverable except in the utmost of thrift and economy both by the individual and by the state. Petty's 'entertainments, magnificent shews, triumphal arches, etc.' gave place to the penny-wisdom of Gladstonian finance and to a state system which 'could not afford' hospitals, open spaces, noble buildings, even the preservation of its ancient monuments, far less the splendours of music and the drama, all of which were consigned to the private charity or magnanimity of improvident individuals.

The doctrine did not reappear in respectable circles for another century, until in the later phase of Malthus the notion of the insufficiency of effective demand takes a definite place as a scientific explanation of unemployment. Since I have already dealt with this somewhat fully in my essay on Malthus,[1] it will be sufficient if I repeat here one or two characteristic passages which I have already quoted in my essay:

> We see in almost every part of the world vast powers of production which are not put into action, and I explain this phenomenon by saying that from the want of a proper distribution of the actual produce adequate motives are not furnished to continued production...I distinctly maintain that an attempt to accumulate very rapidly, which necessarily implies a considerable diminution of unproductive consumption, by greatly impairing the usual motives to production must prematurely check the progress of wealth...But if it be true that an attempt to accumulate very rapidly will occasion such a division between labour and profits as almost to destroy both the motive and the power of future accumulation and consequently the power of maintaining and em-

---

[1] *Essays in Biography*, pp. 139–47 [*JMK*, vol. x, pp. 97–103].

ploying an increasing population, must it not be acknowledged that such an attempt to accumulate, or that saving too much, may be really prejudicial to a country?[1]

The question is whether this stagnation of capital, and subsequent stagnation in the demand for labour arising from increased production without an adequate proportion of unproductive consumption on the part of the landlords and capitalists, could take place without prejudice to the country, without occasioning a less degree both of happiness and wealth than would have occurred if the unproductive consumption of the landlords and capitalists had been so proportioned to the natural surplus of the society as to have continued uninterrupted the motives to production, and prevented first an unnatural demand for labour and then a necessary and sudden diminution of such demand. But if this be so, how can it be said with truth that parsimony, though it may be prejudicial to the producers, cannot be prejudicial to the state; or that an increase of unproductive consumption among landlords and capitalists may not sometimes be the proper remedy for a state of things in which the motives to production fail?[2]

Adam Smith has stated that capitals are increased by parsimony, that every frugal man is a public benefactor, and that the increase of wealth depends upon the balance of produce above consumption. That these propositions are true to a great extent is perfectly unquestionable....But it is quite obvious that they are not true to an indefinite extent, and that the principles of saving, pushed to excess, would destroy the motive to production. If every person were satisfied with the simplest food, the poorest clothing, and the meanest houses, it is certain that no other sort of food, clothing, and lodging would be in existence....The two extremes are obvious; and it follows that there must be some intermediate point, though the resources of political economy may not be able to ascertain it, where, taking into consideration both the power to produce and the will to consume, the encouragement to the increase of wealth is the greatest.[3]

Of all the opinions advanced by able and ingenious men, which I have ever met with, the opinion of M. Say, which

[1] A letter from Malthus to Ricardo, dated 7 July 1821.
[2] A letter from Malthus to Ricardo, dated 16 July 1821.
[3] Preface to Malthus's *Principles of Political Economy*, pp. 8, 9.

states that, *un produit consommé ou détruit est un débouché fermé* (I. i. ch. 15), appears to me to be the most directly opposed to just theory, and the most uniformly contradicted by experience. Yet it directly follows from the new doctrine, that commodities are to be considered only in their relation to each other,—not to the consumers. What, I would ask, would become of the demand for commodities, if all consumption except bread and water were suspended for the next half-year? What an accumulation of commodities! *Quels débouchés!* What a prodigious market would this event occasion![1]

Ricardo, however, was stone-deaf to what Malthus was saying. The last echo of the controversy is to be found in John Stuart Mill's discussion of his wages-fund theory,[2] which in his own mind played a vital part in his rejection of the later phase of Malthus, amidst the discussions of which he had, of course, been brought up. Mill's successors rejected his wages-fund theory but overlooked the fact that Mill's refutation of Malthus depended on it. Their method was to dismiss the problem from the *corpus* of economics not by solving it but by not mentioning it. It altogether disappeared from controversy. Mr Cairncross, searching recently for traces of it amongst the minor Victorians,[3] has found even less, perhaps, than might have been expected.[4] Theories of under-consumption hibernated until the appearance in 1889 of *The Physiology of Industry*, by J. A. Hobson and A. F. Mummery, the first and most significant of many volumes in which for nearly fifty years Mr Hobson has flung himself with unflagging, but almost unavailing, ardour and courage

[1]   Malthus's *Principles of Political Economy*, p. 363, footnote.
[2]   J. S. Mill, *Political Economy*, Book I. chapter v. There is a most important and penetrating discussion of this aspect of Mill's theory in Mummery and Hobson's *Physiology of Industry*, pp. 38 *et seq.*, and, in particular, of his doctrine (which Marshall, in his very unsatisfactory discussion of the wages-fund theory, endeavoured to explain away) that 'a demand for commodities is not a demand for labour'.
[3]   'The Victorians and Investment', *Economic History*, 1936.
[4]   Fullarton's tract *On the Regulation of Currencies* (1844) is the most interesting of his references.

against the ranks of orthodoxy. Though it is so completely forgotten to-day, the publication of this book marks, in a sense, an epoch in economic thought.[1]

The Physiology of Industry was written in collaboration with A. F. Mummery. Mr Hobson has told how the book came to be written as follows:[2]

It was not until the middle 'eighties that my economic heterodoxy began to take shape. Though the Henry George campaign against land values and the early agitation of various socialist groups against the visible oppression of the working classes, coupled with the revelations of the two Booths regarding the poverty of London, made a deep impression on my feelings, they did not destroy my faith in Political Economy. That came from what may be called an accidental contact. While teaching at a school in Exeter I came into personal relations with a business man named Mummery, known then and afterwards as a great mountaineer who had discovered another way up the Matterhorn and who, in 1895, was killed in an attempt to climb the famous Himalayan mountain Nanga Parbat. My intercourse with him, I need hardly say, did not lie on this physical plane. But he was a mental climber as well, with a natural eye for a path of his own finding and a sublime disregard of intellectual authority. This man entangled me in a controversy about excessive saving, which he regarded as responsible for the under-employment of capital and labour in periods of bad trade. For a long time I sought to counter his arguments by the use of the orthodox economic weapons. But at length he convinced me and I went in with him to elaborate the over-saving argument in a book entitled *The Physiology of Industry*, which was published in 1889. This was the first open step in my heretical career, and I did not in the least realise its momentous consequences. For just at that time I had given up my scholastic post and was opening a new line of work as University Extension Lecturer in Economics and Literature. The first shock came in a refusal of the London Extension Board to allow me to

---

[1] J. M. Robertson's *The Fallacy of Saving*, published in 1892, supported the heresy of Mummery and Hobson. But it is not a book of much value or significance, being entirely lacking in the penetrating intuitions of *The Physiology of Industry*.

[2] In an address called 'Confessions of an Economic Heretic', delivered before the London Ethical Society at Conway Hall on Sunday, 14 July 1935. I reproduce it here by Mr Hobson's permission.

offer courses of Political Economy. This was due, I learned, to the intervention of an Economic Professor who had read my book and considered it as equivalent in rationality to an attempt to prove the flatness of the earth. How could there be any limit to the amount of useful saving when every item of saving went to increase the capital structure and the fund for paying wages? Sound economists could not fail to view with horror an argument which sought to check the source of all industrial progress.[1] Another interesting personal experience helped to bring home to me the sense of my iniquity. Though prevented from lecturing on economics in London, I had been allowed by the greater liberality of the Oxford University Extension Movement to address audiences in the Provinces, confining myself to practical issues relating to working-class life. Now it happened at this time that the Charity Organisation Society was planning a lecture campaign upon economic subjects and invited me to prepare a course. I had expressed my willingness to undertake this new lecture work, when suddenly, without explanation, the invitation was withdrawn. Even then I hardly realised that in appearing to question the virtue of unlimited thrift I had committed the unpardonable sin.

In this early work Mr Hobson with his collaborator expressed himself with more direct reference to the classical economics (in which he had been brought up) than in his later writings; and for this reason, as well as because it is the first expression of his theory, I will quote from it to show how significant and well-founded were the authors' criticisms and intuitions. They point out in their preface as follows the nature of the conclusions which they attack:

> Saving enriches and spending impoverishes the community along with the individual, and it may be generally defined as an assertion that the effective love of money is the root of all economic good. Not merely does it enrich the thrifty in

[1] Hobson had written disrespectfully in *The Physiology of Industry*, p. 26: 'Thrift is the source of national wealth, and the more thrifty a nation is the more wealthy it becomes. Such is the common teaching of almost all economists; many of them assume a tone of ethical dignity as they plead the infinite value of thrift; this note alone in all their dreary song has caught the favour of the public ear.'

NOTES ON MERCANTILISM, ETC.

dividual himself, but it raises wages, gives work to the unemployed, and scatters blessings on every side. From the daily papers to the latest economic treatise, from the pulpit to the House of Commons, this conclusion is reiterated and re-stated till it appears positively impious to question it. Yet the educated world, supported by the majority of economic thinkers, up to the publication of Ricardo's work strenuously denied this doctrine, and its ultimate acceptance was exclusively due to their inability to meet the now exploded wages-fund doctrine. That the conclusion should have survived the argument on which it logically stood, can be explained on no other hypothesis than the commanding authority of the great men who asserted it. Economic critics have ventured to attack the theory in detail, but they have shrunk appalled from touching its main conclusions. Our purpose is to show that these conclusions are not tenable, that an undue exercise of the habit of saving is possible, and that such undue exercise impoverishes the Community, throws labourers out of work, drives down wages, and spreads that gloom and prostration through the commercial world which is known as Depression in Trade...

The object of production is to provide 'utilities and conveniences' for consumers, and the process is a continuous one from the first handling of the raw material to the moment when it is finally consumed as a utility or a convenience. The only use of Capital being to aid the production of these utilities and conveniences, the total used will necessarily vary with the total of utilities and conveniences daily or weekly consumed. Now saving, while it increases the existing aggregate of Capital, simultaneously reduces the quantity of utilities and conveniences consumed; any undue exercise of this habit must, therefore, cause an accumulation of Capital in excess of that which is required for use, and this excess will exist in the form of general over-production.[1]

In the last sentence of this passage there appears the root of Hobson's mistake, namely, his supposing that it is a case of excessive saving causing the *actual* accumulation of capital in excess of what is required, which is, in fact, a secondary evil which only occurs through mistakes of foresight; whereas the primary

[1] Hobson and Mummery, *Physiology of Industry*, pp. iii–v.

367   26-2

evil is a propensity to save in conditions of full employment more than the equivalent of the capital which is required, thus preventing full employment except when there is a mistake of foresight. A page or two later, however, he puts one half of the matter, as it seems to me, with absolute precision, though still overlooking the possible rôle of changes in the rate of interest and in the state of business confidence, factors which he presumably takes as given:

> We, are thus brought to the conclusion that the basis on which all economic teaching since Adam Smith has stood, viz. that the quantity annually produced is determined by the aggregates of Natural Agents, Capital, and Labour available, is erroneous, and that, on the contrary, the quantity produced, while it can never exceed the limits imposed by these aggregates, may be, and actually is, reduced far below this maximum by the check that undue saving and the consequent accumulation of over-supply exerts on production; i.e. that in the normal state of modern industrial Communities, consumption limits production and not production consumption.[1]

Finally he notices the bearing of his theory on the validity of the orthodox Free Trade arguments:

> We also note that the charge of commercial imbecility, so freely launched by orthodox economists against our American cousins and other Protectionist Communities, can no longer be maintained by any of the Free Trade arguments hitherto adduced, since all these are based on the assumption that over-supply is impossible.[2]

The subsequent argument is, admittedly, incomplete. But it is the first explicit statement of the fact that capital is brought into existence not by the propensity to save but in response to the demand resulting from actual and prospective consumption. The following portmanteau quotation indicates the line of thought:

> It should be clear that the capital of a community cannot be advantageously increased without a subsequent increase

[1] Hobson and Mummery, *Physiology of Industry*, p. vi.
[2] *Op. cit.* p. ix.

in consumption of commodities...Every increase in
saving and in capital requires, in order to be effectual, a corresponding increase in immediately future consumption[1]...
And when we say future consumption, we do not refer to a
future of ten, twenty, or fifty years hence, but to a future that
is but little removed from the present...If increased thrift
or caution induces people to save more in the present, they
must consent to consume more in the future[2]...No
more capital can economically exist at any point in the productive process than is required to furnish commodities for
the current rate of consumption[3]...It is clear that my
thrift in no wise affects the total economic thrift of the community, but only determines whether a particular portion of
the total thrift shall have been exercised by myself or by
somebody else. We shall show how the thrift of one part
of the community has power to force another part to live
beyond their income.[4]...Most modern economists deny
that consumption could by any possibility be insufficient.
Can we find any economic force at work which might incite
a community to this excess, and if there be any such forces
are there not efficient checks provided by the mechanism of
commerce? It will be shown, firstly, that in every highly
organised industrial society there is constantly at work a force
which naturally operates to induce excess of thrift; secondly,
that the checks alleged to be provided by the mechanism of
commerce are either wholly inoperative or are inadequate to
prevent grave commercial evil[5]...The brief answer
which Ricardo gave to the contentions of Malthus and
Chalmers seems to have been accepted as sufficient by most
later economists. 'Productions are always bought by productions or services; money is only the medium by which
the exchange is effected. Hence the increased production
being always accompanied by a correspondingly increased
ability to get and consume, there is no possibility of Over-
production' (Ricardo, *Prin. of Pol. Econ.* p. 362).[6]

Hobson and Mummery were aware that interest
was nothing whatever except payment for the use of
money.[7] They also knew well enough that their
opponents would claim that there would be 'such a fall

---

[1] *Op. cit.* p. 27.   [2] *Op. cit.* pp. 50, 51.   [3] *Op. cit.* p. 69.
[4] *Op. cit.* p. 113.   [5] *Op. cit.* p. 100.   [6] *Op. cit.* p. 101.
[7] *Op. cit.* p. 79.

in the rate of interest (or profit) as will act as a check upon Saving, and restore the proper relation between production and consumption'.[1] They point out in reply that 'if a fall of Profit is to induce people to save less, it must operate in one of two ways, either by inducing them to spend more or by inducing them to produce less'.[2] As regards the former they argue that when profits fall the aggregate income of the community is reduced, and 'we cannot suppose that when the average rate of incomes is falling, individuals will be induced to increase their rate of consumption by the fact that the premium upon thrift is correspondingly diminished'; whilst as for the second alternative, 'it is so far from being our intention to deny that a fall of profit, due to over-supply, will check production, that the admission of the operation of this check forms the very centre of our argument'.[3] Nevertheless, their theory failed of completeness, essentially on account of their having no independent theory of the rate of interest; with the result that Mr Hobson laid too much emphasis (especially in his later books) on under-consumption leading to over-investment, in the sense of unprofitable investment, instead of explaining that a relatively weak propensity to consume helps to cause unemployment by requiring and *not* receiving the accompaniment of a compensating volume of new investment, which, even if it may sometimes occur temporarily through errors of optimism, is in general prevented from happening at all by the prospective profit falling below the standard set by the rate of interest.

Since the war there has been a spate of heretical theories of under-consumption, of which those of Major Douglas are the most famous. The strength of Major Douglas's advocacy has, of course, largely depended on orthodoxy having no valid reply to much

---

[1] *Op. cit.* p. 117.   [2] *Op. cit.* p. 130.
[3] Hobson and Mummery, *Physiology of Industry*, p. 131.

of his destructive criticism. On the other hand, the detail of his diagnosis, in particular the so-called $A + B$ theorem, includes much mere mystification. If Major Douglas had limited his B-items to the financial provisions made by entrepreneurs to which no current expenditure on replacements and renewals corresponds, he would be nearer the truth. But even in that case it is necessary to allow for the possibility of these provisions being offset by new investment in other directions as well as by increased expenditure on consumption. Major Douglas is entitled to claim, as against some of his orthodox adversaries, that he at least has not been wholly oblivious of the outstanding problem of our economic system. Yet he has scarcely established an equal claim to rank—a private, perhaps, but not a major in the brave army of heretics—with Mandeville, Malthus, Gesell and Hobson, who, following their intuitions, have preferred to see the truth obscurely and imperfectly rather than to maintain error, reached indeed with clearness and consistency and by easy logic but on hypotheses inappropriate to the facts.

Chapter 24

CONCLUDING NOTES ON THE SOCIAL PHILOSOPHY TOWARDS WHICH THE GENERAL THEORY MIGHT LEAD

I

The outstanding faults of the economic society in which we live are its failure to provide for full employment and its arbitrary and inequitable distribution of wealth and incomes. The bearing of the foregoing theory on the first of these is obvious. But there are also two important respects in which it is relevant to the second.

Since the end of the nineteenth century significant progress towards the removal of very great disparities of wealth and income has been achieved through the instrument of direct taxation—income tax and surtax and death duties—especially in Great Britain. Many people would wish to see this process carried much further, but they are deterred by two considerations; partly by the fear of making skilful evasions too much worth while and also of diminishing unduly the motive towards risk-taking, but mainly, I think, by the belief that the growth of capital depends upon the strength of the motive towards individual saving and that for a large proportion of this growth we are dependent on the savings of the rich out of their superfluity. Our argument does not affect the first of these considerations. But it may considerably modify our attitude towards the second. For we have seen that, up to the point where full employment prevails, the growth of capital depends not at all on a low propensity to consume but is, on the contrary, held back by it; and only in conditions of full employment is a low propensity to consume conducive to the growth of capital. Moreover, experience suggests that in existing conditions saving by institutions and through sinking funds is more than adequate, and that measures for the redistribution of incomes in a way likely to raise the propensity to consume may prove positively favourable to the growth of capital.

The existing confusion of the public mind on the matter is well illustrated by the very common belief that the death duties are responsible for a reduction in the capital wealth of the country. Assuming that the State applies the proceeds of these duties to its ordinary outgoings so that taxes on incomes and consumption are correspondingly reduced or avoided, it is, of course, true that a fiscal policy of heavy death duties has the effect of increasing the community's propensity to consume. But inasmuch as an increase in the habitual propensity to consume will in general (i.e. except in conditions of full employment) serve to increase at the same time the inducement to invest, the inference commonly drawn is the exact opposite of the truth.

Thus our argument leads towards the conclusion that in contemporary conditions the growth of wealth, so far from being dependent on the abstinence of the rich, as is commonly supposed, is more likely to be impeded by it. One of the chief social justifications of great inequality of wealth is, therefore, removed. I am not saying that there are no other reasons, unaffected by our theory, capable of justifying some measure of inequality in some circumstances. But it does dispose of the most important of the reasons why hitherto we have thought it prudent to move carefully. This particularly affects our attitude towards death duties: for there are certain justifications for inequality of incomes which do not apply equally to inequality of inheritances.

For my own part, I believe that there is social and psychological justification for significant inequalities of incomes and wealth, but not for such large disparities as exist to-day. There are valuable human activities which require the motive of money-making and the environment of private wealth-ownership for their full fruition. Moreover, dangerous human proclivities can be canalised into comparatively harmless channels by the existence of opportunities for money-making and private wealth, which, if they cannot be satisfied in this way, may find their outlet in cruelty, the reckless pursuit of personal power and authority, and other forms of self-aggrandisement. It is better that a man should tyrannise over his bank balance than over his fellow-citizens; and whilst the former is sometimes denounced as being but a means to the latter, sometimes at least it is an alternative. But it is not necessary for the stimulation of these activities and the satisfaction of these proclivities that the game should be played for such high stakes as at present. Much lower stakes will serve the purpose equally well, as soon as the players are accustomed to them. The task of transmuting human nature must not be confused with the task of managing it. Though in the ideal commonwealth men may have been taught or inspired or bred to take no interest in the stakes, it may still be wise and prudent statesmanship to allow the game to be played, subject to rules and limitations, so long as the average man, or even a significant section of the community, is in fact strongly addicted to the money-making passion.

II

There is, however, a second, much more fundamental inference from our argument which has a bearing on the future of inequalities of wealth; namely, our theory of the rate of interest. The justification for a moderately high rate of interest has been found hitherto in the necessity of providing a sufficient inducement to save. But we have shown that the extent of effective saving is necessarily determined by the scale of investment and that the scale of investment is promoted by a *low* rate of interest, provided that we do not attempt to stimulate it in this way beyond the point which corresponds to full employment. Thus it is to our best advantage to reduce the rate of interest to that point relatively to the schedule of the marginal efficiency of capital at which there is full employment.

There can be no doubt that this criterion will lead to a much lower rate of interest than has ruled hitherto; and, so far as one can guess at the schedules of the marginal efficiency of capital corresponding to increasing amounts of capital, the rate of interest is likely to fall steadily, if it should be practicable to maintain conditions of more or less continuous full employment—unless, indeed, there is an excessive change in the aggregate propensity to consume (including the State).

I feel sure that the demand for capital is strictly limited in the sense that it would not be difficult to increase the stock of capital up to a point where its marginal efficiency had fallen to a very low figure. This would not mean that the use of capital instruments would cost almost nothing, but only that the return from them would have to cover little more than their exhaustion by wastage and obsolescence together with some margin to cover risk and the exercise of skill and judgment. In short, the aggregate return from durable goods in the course of their life would, as in the case of short-lived goods, just cover their labour-costs of production *plus* an allowance for risk and the costs of skill and supervision.

Now, though this state of affairs would be quite compatible with some measure of individualism, yet

375

it would mean the euthanasia of the rentier, and, consequently, the euthanasia of the cumulative oppressive power of the capitalist to exploit the scarcity-value of capital. Interest to-day rewards no genuine sacrifice, any more than does the rent of land. The owner of capital can obtain interest because capital is scarce, just as the owner of land can obtain rent because land is scarce. But whilst there may be intrinsic reasons for the scarcity of land, there are no intrinsic reasons for the scarcity of capital. An intrinsic reason for such scarcity, in the sense of a genuine sacrifice which could only be called forth by the offer of a reward in the shape of interest, would not exist, in the long run, except in the event of the individual propensity to consume proving to be of such a character that net saving in conditions of full employment comes to an end before capital has become sufficiently abundant. But even so, it will still be possible for communal saving through the agency of the State to be maintained at a level which will allow the growth of capital up to the point where it ceases to be scarce.

I see, therefore, the rentier aspect of capitalism as a transitional phase which will disappear when it has done its work. And with the disappearance of its rentier aspect much else in it besides will suffer a sea-change. It will be, moreover, a great advantage of the order of events which I am advocating, that the euthanasia of the rentier, of the functionless investor, will be nothing sudden, merely a gradual but prolonged continuance of what we have seen recently in Great Britain, and will need no revolution.

Thus we might aim in practice (there being nothing in this which is unattainable) at an increase in the volume of capital until it ceases to be scarce, so that the functionless investor will no longer receive a bonus; and at a scheme of direct taxation which allows the intelligence and determination and executive skill of the financier, the entrepreneur *et hoc genus omne* (who are certainly so fond of their craft that their labour could be obtained much cheaper than at present), to be harnessed to the service of the community on reasonable terms of reward.

At the same time we must recognise that only experience can show how far the common will, embodied in the policy of the State, ought to be directed to increasing and supplementing the inducement to invest; and how far it is safe to stimulate the average propensity to consume, without foregoing our aim of depriving capital of its scarcity-value within one or two generations. It may turn out that the propensity to consume will be so easily strengthened by the effects of a falling rate of interest, that full employment can be reached with a rate of accumulation little greater than at present. In this event a scheme for the higher taxation of large incomes and inheritances might be open to the objection that it would lead to full employment with a rate of accumulation which was reduced considerably below the current level. I must not be supposed to deny the possibility, or even the probability, of this outcome. For in such matters it is rash to predict how the average man will react to a changed environment. If, however, it should prove easy to secure an approximation to full employment with a rate of accumulation not much greater than at present, an outstanding problem will at least have been solved. And it would remain for separate decision on what scale and by what means it is right and reasonable to call on the living generation to restrict their consumption, so as to establish in course of time, a state of full investment for their successors.

III

In some other respects the foregoing theory is moderately conservative in its implications. For whilst it indicates the vital importance of establishing certain central controls in matters which are now left in the main to individual initiative, there are wide fields of activity which are unaffected. The State will have to exercise a guiding influence on the propensity to consume partly through its scheme of taxation, partly by fixing the rate of interest, and partly, perhaps, in other ways. Furthermore, it seems unlikely that the influence of banking policy on the rate of interest will be sufficient by itself to determine an optimum rate of investment. I conceive, therefore, that a somewhat comprehensive socialisation of investment will prove the only means of securing an approximation to full employment; though this need not exclude all manner of compromises and of devices by which public authority will co-operate with private initiative. But beyond this no obvious case is made out for a system of State Socialism which would embrace most of the economic life of the community. It is not the ownership of the instruments of production which it is important for the State to assume. If the State is able to determine the aggregate amount of resources devoted to augmenting the instruments and the basic rate of reward to those who own them, it will have accomplished all that is necessary. Moreover, the necessary measures of socialisation can be introduced gradually and without a break in the general traditions of society.

Our criticism of the accepted classical theory of economics has consisted not so much in finding logical flaws in its analysis as in pointing out that its tacit assumptions are seldom or never satisfied, with the result that it cannot solve the economic problems of the actual world. But if our central controls succeed in establishing an aggregate volume of output corresponding to full employment as nearly as is practicable, the classical theory comes into its own again from this point onwards. If we suppose the volume of output to be given, i.e. to be determined by forces outside the classical scheme of thought, then there is no objection to be raised against the classical analysis of the manner in which private self-interest will determine what in particular is produced, in what proportions the factors of production will be combined to produce it, and how the value of the final product will be distributed between them. Again, if we have dealt otherwise with the problem of thrift, there is no objection to be raised against the modern classical theory as to the degree of consilience between private and public advantage in conditions of perfect and imperfect competition respectively. Thus, apart from the necessity of central controls to bring about an adjustment between the propensity to consume and the inducement to invest, there is no more reason to socialise economic life than there was before.

To put the point concretely, I see no reason to suppose that the existing system seriously misemploys the factors of production which are in use. There are, of course, errors of foresight; but these would not be avoided by centralising decisions. When 9,000,000 men are employed out of 10,000,000 willing and able to work, there is no evidence that the labour of these 9,000,000 men is misdirected. The complaint against the present system is not that these 9,000,000 men ought to be employed on different tasks, but that tasks should be available for the remaining 1,000,000 men. It is in determining the volume, not the direction, of actual employment that the existing system has broken down.

Thus I agree with Gesell that the result of filling in the gaps in the classical theory is not to dispose of the 'Manchester System', but to indicate the nature of the environment which the free play of economic forces requires if it is to realise the full potentialities of production. The central controls necessary to ensure full employment will, of course, involve a large extension of the traditional functions of government. Furthermore, the modern classical theory has itself called attention to various conditions in which the free play of

379

economic forces may need to be curbed or guided. But there will still remain a wide field for the exercise of private initiative and responsibility. Within this field the traditional advantages of individualism will still hold good.

Let us stop for a moment to remind ourselves what these advantages are. They are partly advantages of efficiency—the advantages of decentralisation and of the play of self-interest. The advantage to efficiency of the decentralisation of decisions and of individual responsibility is even greater, perhaps, than the nineteenth century supposed; and the reaction against the appeal to self-interest may have gone too far. But, above all, individualism, if it can be purged of its defects and its abuses, is the best safeguard of personal liberty in the sense that, compared with any other system, it greatly widens the field for the exercise of personal choice. It is also the best safeguard of the variety of life, which emerges precisely from this extended field of personal choice, and the loss of which is the greatest of all the losses of the homogeneous or totalitarian state. For this variety preserves the traditions which embody the most secure and successful choices of former generations; it colours the present with the diversification of its fancy; and, being the handmaid of experiment as well as of tradition and of fancy, it is the most powerful instrument to better the future.

Whilst, therefore, the enlargement of the functions of government, involved in the task of adjusting to one another the propensity to consume and the inducement to invest, would seem to a nineteenth-century publicist or to a contemporary American financier to be a terrific encroachment on individualism, I defend it, on the contrary, both as the only practicable means of avoiding the destruction of existing economic forms in their entirety and as the condition of the successful functioning of individual initiative.

For if effective demand is deficient, not only is the

public scandal of wasted resources intolerable, but the individual enterpriser who seeks to bring these resources into action is operating with the odds loaded against him. The game of hazard which he plays is furnished with many zeros, so that the players *as a whole* will lose if they have the energy and hope to deal all the cards. Hitherto the increment of the world's wealth has fallen short of the aggregate of positive individual savings; and the difference has been made up by the losses of those whose courage and initiative have not been supplemented by exceptional skill or unusual good fortune. But if effective demand is adequate, average skill and average good fortune will be enough.

The authoritarian state systems of to-day seem to solve the problem of unemployment at the expense of efficiency and of freedom. It is certain that the world will not much longer tolerate the unemployment which, apart from brief intervals of excitement, is associated—and, in my opinion, inevitably associated—with present-day capitalistic individualism. But it may be possible by a right analysis of the problem to cure the disease whilst preserving efficiency and freedom.

IV

I have mentioned in passing that the new system might be more favourable to peace than the old has been. It is worth while to repeat and emphasise that aspect.

War has several causes. Dictators and others such, to whom war offers, in expectation at least, a pleasurable excitement, find it easy to work on the natural bellicosity of their peoples. But, over and above this, facilitating their task of fanning the popular flame, are the economic causes of war, namely, the pressure of population and the competitive struggle for markets. It is the second factor, which probably played a predominant part in the nineteenth century, and might again, that is germane to this discussion.

I have pointed out in the preceding chapter that, under the system of domestic *laissez-faire* and an international gold standard such as was orthodox in the latter half of the nineteenth century, there was no means open to a government whereby to mitigate economic distress at home except through the competitive struggle for markets. For all measures helpful to a state of chronic or intermittent under-employment were ruled out, except measures to improve the balance of trade on income account.

Thus, whilst economists were accustomed to applaud the prevailing international system as furnishing the fruits of the international division of labour and harmonising at the same time the interests of different nations, there lay concealed a less benign influence; and those statesmen were moved by common sense and a correct apprehension of the true course of events, who believed that if a rich, old country were to neglect the struggle for markets its prosperity would droop and fail. But if nations can learn to provide themselves with full employment by their domestic policy (and, we must add, if they can also attain equilibrium in the trend of their population), there need be no important economic forces calculated to set the interest of one country against that of its neighbours. There would still be room for the international division of labour and for international lending in appropriate conditions. But there would no longer be a pressing motive why one country need force its wares on another or repulse the offerings of its neighbour, not because this was necessary to enable it to pay for what it wished to purchase, but with the express object of upsetting the equilibrium of payments so as to develop a balance of trade in its own favour. International trade would cease to be what it is, namely, a desperate expedient to maintain employment at home by forcing sales on foreign markets and restricting purchases, which, if successful, will merely shift the problem of unemployment to the neighbour which is worsted in the struggle, but a willing and unimpeded exchange of goods and services in conditions of mutual advantage.

V

Is the fulfilment of these ideas a visionary hope? Have they·insufficient roots in the motives which govern the evolution of political society? Are the interests which they will thwart stronger and more obvious than those which they will serve?

I do not attempt an answer in this place. It would need a volume of a different character from this one to indicate even in outline the practical measures in which they might be gradually clothed. But if the ideas are correct—an hypothesis on which the author himself must necessarily base what he writes—it would be a mistake, I predict, to dispute their potency over a period of time. At the present moment people are unusually expectant of a more fundamental diagnosis; more particularly ready to receive it; eager to try it out, if it should be even plausible. But apart from this contemporary mood, the ideas of economists and political philosophers, both when they are right and when they are wrong, are more powerful than is commonly understood. Indeed the world is ruled by little else. Practical men, who believe themselves to be quite exempt from any intellectual influences, are usually the slaves of some defunct economist. Madmen in authority, who hear voices in the air, are distilling their frenzy from some academic scribbler of a few years back. I am sure that the power of vested interests is vastly exaggerated compared with the gradual encroachment of ideas. Not, indeed, immediately, but after a certain interval; for in the field of economic and political philosophy there are not many who are

383

influenced by new theories after they are twenty-five or thirty years of age, so that the ideas which civil servants and politicians and even agitators apply to current events are not likely to be the newest. But, soon or late, it is ideas, not vested interests, which are dangerous for good or evil.

Appendix 1

Page	Line	Correction
60	6	For 'possession' read 'possessions'
83	12	For 'has' read 'had'
126	13	For '23' read '19'
126 footnote 1 line 2		For 'th' read 'the'
172	21	For 'security' read 'precautionary'
212	9	For 'than' read 'that'
229	32	For 'output' read 'the stock of assets in general'
233	25	For 'their' read 'its'
237	31	For 'or' read 'of'
267	28	For 'three' read 'four'
271	4	For 'technique' read 'techniques'
319	23	For 'income' read 'incomes'
341	7	For 'Mercantilist' read 'Mercantilists'

These corrections come to light in preparing various foreign editions of *The General Theory*, in preparing the variorum version of earlier drafts which appears in volume XIV, or in setting this book for press. The corrections do not cover more substantial errors such as the unsatisfactory presentation of aggregate supply and demand on page 29 or the inadequate derivation of the equations on page 305.

Appendix 2

From *The Economic Journal*, September 1936

FLUCTUATIONS IN NET INVESTMENT IN THE UNITED STATES

In my *General Theory of Employment, Interest and Money*, chap. 8, pp. 98–104, I made a brief attempt to illustrate the wide range of fluctuations in net investment, basing myself on certain calculations by Mr Colin Clark for Great Britain and by Mr Kuznets for the United States.[1]

In the case of Mr Kuznets' figures I pointed out (p. 103) that his allowances for depreciation, etc., included 'no deduction at all in respect of houses and other durable commodities in the hands of individuals'. But the table which immediately followed this did not make it sufficiently clear to the reader that the first line relating to 'gross capital formation' comprised much wider categories of capital goods than the second line relating to 'entrepreneurs' depreciation, etc.'; and I was myself misled on the next page, where I expressed doubts as to the sufficiency of the latter item in relation to the former (forgetting that the latter related only to a part of the former). The result was that the table as printed considerably under-stated the force of the phenomenon which I was concerned to describe, since a complete calculation in respect of depreciation, etc., covering all the items in the first line of the table, would lead to much larger figures than those given in the second line. Some correspondence with Mr Kuznets now enables me to explain these important figures more fully and clearly, and in the light of later information.

Mr Kuznets divides his aggregate of gross capital formation (as he calls it) for the United States into a number of categories as follows:

(1) *Consumers' Durable Commodities*

These comprise motor-cars, furniture and house equipment and other more or less durable articles, apart from houses, pur-

[1] Published in *Bulletin* 52, 15 November 1934, of the National Bureau of Economic Research (New York).

chased and owned by those who consume them. Whether or not these items should be included in investment depends (so far as the definition is concerned) on whether the expenditure on them when it is initially made is included in current saving or in current expenditure; and it depends (so far as the practical application is concerned) on whether in subsequent years the owners feel under a motive to make provision for current depreciation out of their incomes even when they are not replacing or renewing them. Doubtless it is not possible to draw a hard-and-fast line. But it is probable that few individuals feel it necessary in such cases to make a financial provision for depreciation apart from actual repairs and renewals. This, in combination with the difficulty of obtaining proper statistics and of drawing a clear line, makes it preferable, I think, to exclude such equipment from investment and to include it in consumption-expenditure in the year in which it is incurred. This is in accordance with the definition of *consumption* given in my *General Theory*, p. 54.[1]

I shall, therefore, exclude this category from the final calculation;[2] though I hope to deal with the problem more thoroughly at a later time. Nevertheless it may be interesting to quote Mr Kuznets' estimates, which are of substantial magnitude:

	(Millions of dollars.)								
	1925	1926	1927	1928	1929	1930	1931	1932	1933
Consumers' durable commodities	8,664	9,316	8,887	9,175	10,058	7,892	5,885	4,022	3,737

The above figure for 1929 includes 3,400 million dollars for motor-cars, whilst the depreciation in respect of the same item for that year is estimated at 2,500 million dollars.

(2) *Residential Construction*

This is an important and highly fluctuating item which should undoubtedly be included in investment, and not in consumption expenditure, since houses are usually regarded as

[1] See also *op. cit.*, pp. 61, 62, where I should have made it clearer that the purchase of a house is most conveniently regarded as an act of entrepreneurship.

[2] It was (inconsistently) included in the figures for gross capital formation given in my *General Theory*, p. 103.

purchased out of savings and not out of income, and are often owned by others than the occupiers. In the *Bulletin* from which these figures are taken Mr Kuznets gives no estimate for the annual rate of depreciation, etc. More recently, however, his colleague, Mr Solomon Fabricant, has published such estimates,[1] which I have used in the following table:

				(Millions of dollars.)					
	1925	1926	1927	1928	1929	1930	1931	1932	1933
Residential construction	3,050	2,965	2,856	3,095	2,127	1,222	900	311	276
Depreciation*	1,554	1,676	1,754	1,842	1,911	1,901	1,698	1,460	1,567
Net investment	1,496	1,289	1,102	1,253	216	−679	−798	−1,149	−1,291

* These figures are calculated in terms of current (reproduction) costs. Mr Fabricant has also provided estimates in terms of original cost, which for the years prior to 1932 are considerably lower.

(3) *Business Fixed Capital*

Mr Kuznets here distinguishes expenditure on new producers' durable goods and business construction from the net change in 'business inventories,' i.e. in working and liquid capital; and we shall, therefore, deal with the latter under a separate heading.

The amount of the deduction to obtain net investment in respect of parts, repairs and servicing, and repairs and maintenance of business construction as distinct from depreciation and depletion, which is not made good, depends, of course, on whether the former have been included in gross investment. Mr Kuznets gives a partial estimate for the former but the figures given below exclude these items both from gross and from net investment. But whilst the result of deducting both the repairs item and the depreciation item probably corresponds fairly closely to my *net investment*, the two deductions taken separately do not closely correspond to my deductions for *user cost* and *supplementary cost*; so that it is not possible to calculate from Mr Kuznets' *data* a figure corresponding to my (gross) *investment*.

The following table gives in the first line 'the formation of gross capital destined for business use, exclusive of parts, repairs and servicing, and repairs and maintenance of business construction, and excluding changes in business inventories';

[1] 'Measures of Capital Consumption (1919–33)', National Bureau of Economic Research, *Bulletin* 60.

and in the second line the estimated 'depreciation and depletion' on the same items:

				(Millions of dollars.)					
	1925	1926	1927	1928	1929	1930	1931	1932	1933
Gross business capital formation (as above)	9,070	9,815	9,555	10,019	11,396	9,336	5,933	3,205	2,894
Depreciation and depletion*	5,685	6,269	6,312	6,447	7,039	6,712	6,154	5,092	4,971
Net investment	3,385	3,546	3,243	3,572	4,357	2,624	−221	−1,887	−2,077

* These figures are not taken from Mr Kuznets' memoranda, but from Mr Fabricant's later and revised estimates. As before they are in terms of current (replacement) cost. In terms of original cost they are appreciably lower prior to 1931 and higher subsequently.

(4) *Business Inventories*

For the financial gains or losses arising out of this item there appear to be fairly adequate statistics in the United States, though not in this country. Mr Kuznets' figures are as follows:

				(Millions of dollars.)					
	1925	1926	1927	1928	1929	1930	1931	1932	1933
Net gain or loss in business inventories	916	2,664	−176	511	1,800	100	−500	−2,250	−2,250

This table covers not only manufacturers' stocks but also stocks of farmers, mines, traders, government agencies, etc. From 1929 onwards the figures given in Mr Kuznets' memorandum of 1934 proved to require correction. Those given above are provisional and approximate estimates, pending the publication of revised figures by the National Bureau.

(5) *Public Construction and Borrowing*

The relevant figure in this context is not so much the gross (or net) expenditure on construction, as the amount of expenditure met out of a net increase in borrowing. That is to say in the case of public authorities and the like, their net investment may be best regarded as being measured by the net increase in their borrowing. In so far as their expenditures are met by compulsory transfer from the current income of the public, they have no correlative in private saving; whilst public saving, if we were to find a satisfactory definition for this concept, would be subject to quite different psychological influences from private saving. I have touched on the problem in my *General Theory*, p. 128, footnote. I propose, therefore, to insert in place of the figures of public construction the 'loan expenditure' of public bodies.

Mr Kuznets has very kindly supplied me with figures for the net changes in the amount of public debt (Federal, State and local) outstanding in the United States, which, except for minor changes in the Government's cash balances, represent the amount of public expenditure not covered by taxes and other revenues.[1] This is given below in parallel with his estimates of the amount of construction by public authorities. The interesting result emerges that up to 1928 there was a net reduction in the public debt in spite of a large expenditure on public construction, and that even up to 1931 some part of public construction was met out of revenue. The excess of borrowing over construction in 1932 and 1933 represents, of course, various measures of public relief.

	(Millions of dollars.)								
	1925	1926	1927	1928	1929	1930	1931	1932	1933
Public construction*	2,717	2,612	3,045	3,023	2,776	3,300	2,906	2,097	1,659
Net change in outstanding public debt†	−43	−280	−244	−10	+441	+1,712	+2,822	+2,565	+2,796

* See Mr Kuznets' Bulletin, Table II, line 22, brought up to date on the basis of more recent data.
† See col. 9 of the table given in the appendix below.

(6) *Foreign Investment*

Finally, we have the net change in claims against foreign countries, estimated by Mr Kuznets as follows:

	(Millions of dollars.)								
1925	1926	1927	1928	1929	1930	1931	1932	1933
428	44	606	957	312	371	326	40	293

(7) *Aggregate Net Investment*

We are now in a position to combine the above items into a single aggregate. This total is not quite comprehensive, since it excludes construction by semi-public agencies, and a small

[1] The details of Mr Kuznets' compilation are given in a note below.

amount unallocable construction. But Mr Kuznets is of the opinion that both omissions are quite minor in character and could not much affect the movements of net investment in the table which now follows.

	(Millions of dollars.)								
	1925	1926	1927	1928	1929	1930	1931	1932	1933 .
Residential construction	1,496	1,289	1,102	1,253	216	−679	−798	−1,149	−1,291
Business fixed capital	3,385	3,546	3,243	3,572	4,357	2,624	−221	−1,887	−2,077
Business inventories	916	2,664	−176	511	1,800	100	−500	−2,250	−2,250
Net loan expenditures by public authorities	−43	−280	−244	−10	441	1,712	2,822	2,565	2,796
Foreign investment	428	44	606	957	312	371	326	40	293
Aggregate net investment	6,182	7,263	4,531	6,283	7,126	4,128	1,629	−2,681	−2,529

It is evident that this table is of first-class importance for the interpretation of business fluctuations in the United States. In matters of detail the following points stand out:

(a) The arrears of residential construction at the end of 1933 must have been enormous. For there had been no net investment in this field since 1925. This does not mean, of course, that the actual state of housing was so bad as this. Some gross investment in housing continued throughout, and the gradual deterioration in the state of accommodation, through obsolescence and decay not made good, does not impair forthwith to an equal extent the actual accommodation available for the time being.

(b) The part played by fluctuations in business inventories is very marked, especially in accentuating the depression at the bottom of the slump. The increase in inventories in 1929 was probably for the most part designed to meet demand which did not fully materialise; whilst the small further increase in 1930 represented accumulations of unsold stocks. In 1932 and 1933, manufacturers met current demand to an extraordinary extent out of stocks, so that effective demand fell largely behind actual consumption. But this, fortunately, is a state of affairs which could not continue indefinitely. A further depletion of stocks on this scale could not possibly take place, since the stocks were no longer there. A level of business inventories so low as that which existed in the United States at the end of 1933 was an almost certain herald of some measure of recovery. In general an aggregate of net investment which is based on an increase in business inventories beyond normal is clearly precarious; and it is easy to see in retrospect that a large growth of inventories in 1929, coupled with a decline in residential construction, was ominous. The figures for 1934, 1935, and 1936 will be most interesting when we have them. One would expect that the recovery of the two former years has been based on a return of inventories to normal and on public loan expenditure, but that by 1936 durable investment was beginning to supplant inventories in making up the total. It is on the continued steadiness of the first two items of the above table at figures not less than those of 1925 to 1928 that the maintenance of prosperity must depend; and it is for this reason that a low long-term rate of interest is so vitally important.

(c) The manner in which the changes in public loan expenditure came in to moderate the fluctuations, which would have occurred otherwise, is very apparent. The manner in which from 1931 Federal borrowing took the place of State and local borrowing, as shown in the Appendix below, is striking. From 30 June 1924, to 30 June 1930, Federal loans outstanding fell from 21 to 15 billions, whilst in the same period State and local loans rose from 10 to 16 billions, the total remaining unchanged; whereas from 30 June 1930 to 30 June 1935, Federal loans rose from 15 to 26 billions and the others from 16 only to 17 billions. The appendix, which gives the figures of public borrowing up to 30 June 1935, shows—contrary, perhaps, to the general impression— that public borrowing was at its height in 1931, and that in 1934–35 it was but little more than in 1929–30.

(d) When comparable figures of income are available, we shall be able to make some computations as to the value of the Multiplier in the conditions of the United States, though there are many statistical difficulties still to overcome. If, however, as a very crude, preliminary test we take the Dept. of Commerce estimates of income (uncorrected for price changes), we find that during the large movements of the years from 1929 to 1932 the changes in money-incomes were from three to five times the changes in net investment shown above. In 1933 incomes and investment both increased slightly, but the movements were too narrow to allow the ratio of the one to the other to be calculated within a reasonable margin of error.

J. M. KEYNES

A Note to Appendix 2

TOTAL AND NET OUTSTANDING ISSUES OF PUBLIC DEBT

(Millions of dollars.)

Date (30 June) (1)	Total outstanding issues			Net outstanding issues			Net change (8)	Average for calendar year (9)
	Federal (2)	State, county, city, etc. (3)	Com- bined (4)	Federal (5)	State, county, city, etc. (6)	Com- bined (7)		
1924	20,982	11,633	32,615	20,627	9,921	30,548	—	—
1925	20,211	12,830	33,041	19,737	10,975	30,712	+164	−43
1926	19,384	13,664	33,048	18,790	11,672	30,462	−250	−280
1927	18,251	14,735	32,986	17,542	12,610	30,152	−310	−244
1928	17,318	15,699	33,017	16,522	13,452	29,974	−178	−10
1929	16,639	16,760	33,399	15,773	14,358	30,131	+157	+441
1930	15,922	17,985	33,907	14,969	15,887	30,856	+725	+1,712
1931	16,520	19,188	35,708	16,098	17,457	33,555	+2,699	+2,822
1932	19,161	19,635	38,796	18,673	17,828	36,501	+2,946	+2,565
1933	22,158	19,107	41,265	21,613	17,072	38,685	+2,184	+2,796
1934	26,480	18,942	45,422	25,323	16,771	42,094	+3,409	+2,173
1935	27,645	19,277	46,922	26,137	16,895	43,032	+938	—

(*Source*: Report of the Secretary of the Treasury for year ended 30 June 1935, p. 424.)

Total outstanding issues exclude a small volume of matured and non-interest bearing obligations (see *ibid.*, p. 379).

Net outstanding issues are equal to total outstanding issues less those held in U.S. Government trust funds, or owned by U.S. Government or by governmental agencies and held in sinking funds.

The table above does not include the contingent debt of the Federal Government, i.e. obligations guaranteed by the United States. These, comprising largely debt issues of the Federal Farm Mortgage Corporation, Home Owners Loan Corporation and the Reconstruction Finance Corporation, were as follows:

Date	Millions of dollars
30 June 1934	691
31 December 1934	3,079
30 June 1935	4,151
31 December 1935	4,525

(See *Cost of Government in the United States*, by the National Industrial Conference Board, pub. no. 223, New York, 1936, Table 26, p. 68.)

Appendix 3

From *The Economic Journal*, March 1939

RELATIVE MOVEMENTS OF REAL WAGES AND OUTPUT

An article by Mr J. G. Dunlop in this *Journal* (September 1938, Vol. XLVIII, p. 413) on *The Movement of Real and Money Wage Rates*, and the note by Mr L. Tarshis printed below [in the *Economic Journal*, March 1939] (p. 150),[1] clearly indicate that a common belief to which I acceded in my *General Theory of Employment* needs to be reconsidered. I there said:

> It would be interesting to see the results of a statistical enquiry into the actual relationship between changes in money wages and changes in real wages. In the case of a change peculiar to a particular industry one would expect the change in real wages to be in the same direction as the change in money wages. But in the case of changes in the general level of wages, it will be found, I think, that the change in real wages associated with a change in money wages, so far from being usually in the same direction, is almost always in the opposite direction...This is because, in the short period, falling money wages and rising real wages are each, for independent reasons, likely to accompany decreasing employment; labour being readier to accept wage-cuts when employment is falling off, yet real wages inevitably rising in the same circumstances on account of the increasing marginal return to a given capital equipment when output is diminished.

But Mr Dunlop's investigations into the British statistics appear to show that, when money wages are rising, real wages have usually risen also; whilst, when money wages are falling, real wages are no more likely to rise than to fall. And Mr Tarshis has reached broadly similar results in respect of recent years in the United States.

In the passage quoted above from my *General Theory* I was accepting, without taking care to check the facts for myself, a belief which has been widely held by British economists up to the last year or two. Since the material on which Mr Dunlop mainly depends—namely, the indices of real and money wages

[1] Cf. also his article on 'Real Wages in the United States and Great Britain', published in *The Canadian Journal of Economics* for August 1938.

prepared by Mr G. H. Wood and Prof. Bowley—have been available to all of us for many years, it is strange that the correction has not been made before.[1] But the underlying problem is not simple, and is not completely disposed of by the statistical studies in question.

First of all it is necessary to distinguish between two different problems. In the passage quoted above I was dealing with the reaction of real wages to changes in *output*, and had in mind situations where changes in real and money wages were a reflection of changes in the level of employment caused by changes in effective demand. This is, in fact, the case which, if I understand them rightly, Mr Dunlop and Mr Tarshis have primarily in view.[2] But there is also the case where changes in wages reflect changes in prices or in the conditions governing the wage bargain which do not correspond to, or are not primarily the result of, changes in the level of output and employment and are not caused by (though they may cause) changes in effective demand. This question I discussed in a different part of my *General Theory* (namely chapter 19, 'Changes in Money Wages'), where I reached the conclusion that wage changes, which are not in the first instance due to changes in output, have complex reactions on output which may be in either direction according to circumstances and about which it is difficult to generalise. It is with the first problem only that I am concerned in what follows.[3]

The question of the influence on real wages of periods of boom and depression has a long history. But we need not go farther back than the period of the 'eighties and 'nineties of the last century, when it was the subject of investigation by various official bodies before which Marshall gave evidence or in the work of which he took part. I was myself brought up upon the evidence he gave before the Gold and Silver Commission in 1887 and the Indian Currency Committee in 1899.[4] It is not always clear

---

[1] Cf., however, the reference given below (p. 399) to Prof. Pigou's *Industrial Fluctuations*.

[2] See, however, the *post-scriptum* to Mr Tarshis's note to which I refer further below.

[3] In his *Essays in the Theory of Economic Fluctuations*, to which I shall have occasion to refer below, Dr Kalecki deals with the relation between real wages and output in the essay entitled 'The Distribution of the National Income.' But it is with the other problem that he is primarily concerned in the essay entitled 'Money and Real Wages.'

[4] Marshall's contributions to official inquiries from 1886 to 1903 we used to regard as constituting, together with the *Principles*, his most important

whether Marshall has in mind a rise in money wages associated with a rise in output, or one which merely reflects a change in prices (due, for example, to a change in the standard which was the particular subject on which he was giving evidence); but in some passages it is evident that he is dealing with changes in real wages at times when output is expanding. It *is* clear, however, that his conclusion is based, not like some later arguments on *à priori* grounds arising out of increasing marginal cost in the short period, but on statistical grounds which showed—so he thought—that in the short period wages were stickier than prices. In his preliminary memorandum for the Gold and Silver Commission (*Official Papers*, p. 19) he wrote: '[During a slow and gradual fall of prices] a powerful friction tends to prevent money wages in most trades from falling as fast as prices; and this tends almost imperceptibly to establish a higher standard of living among the working classes, and to diminish the inequalities of wealth. These benefits are often ignored; but in my opinion they are often nearly as important as the evils which result from that gradual fall of prices which is sometimes called a depression of trade.' And when Mr Chaplin asked him (*op. cit.*, p. 99), 'You think that during a period of depression the employed working classes have been getting more than they did before?' he replied, 'More than they did before, on the average.'

Subsequently, as appears from an important letter of April 1897 (hitherto unpublished) to Foxwell,[1] who held somewhat strongly the opposite opinion, Marshall's opinion became rather more tentative; though the following extract refers more to his general attitude towards rising prices than to their particular effect on real wages:

> You know, my views on this matter are (*a*) not very confident, (*b*) not very warmly advocated by me, (*c*) not very old, (*d*) based entirely on non-academic arguments & observation.
> In the years 68 to 77 I was strongly on the side you now advocate. The observation of events in Bristol made me doubt. In 85, or 86 I wrote a Mem^n for the Com^n on

and valuable work. Re-reading his *Official Papers* to-day, I find this confirmed. Yet his *Official Papers*, published by the Royal Economic Society in 1926 (still obtainable by members at 5*s*.), has had a negligible circulation compared with any of his other works.

[1] Endorsed by Foxwell—'*Marshall*, a very characteristic letter on the question of rising and falling prices, among other matters.'

Depression showing a slight preference for rising prices. But in the following two years I studied the matter closely, I read and analysed the evidence of business men before that Commission; & by the time the Gold & Silver Commission came, I had just turned the corner.

Since then I have read a great deal, but almost exclusively of a non-academic order on the subject: & was thinking about it during a great part of the evidence given by business men & working men before the Labour Commission. I have found a good deal that is new to strengthen my new conviction, nothing to shake it. I am far from certain I am right. I am absolutely certain that the evidence brought forward in print to the contrary in England and America (I have not read largely for other countries) does not prove what it claims to, & does not meet or anticipate my arguments, in the simple way you seem to imagine.

Shortly afterwards he began to work at his evidence for the Indian Currency Committee which seems to have had the effect of confirming him in his previous opinion. His final considered opinion is given in Question 11,781:[1]

I will confess that, for ten or fifteen years after I began to study political economy, I held the common doctrine, that a rise of prices was generally beneficial to business men directly, and indirectly to the working classes. But, after that time, I changed my views, and I have been confirmed in my new opinions by finding that they are largely held in America, which has recently passed through experiences somewhat similar to those of England early in the century. The reasons for the change in my opinion are rather long, and I gave them at some length before the Gold and Silver Commission. I think, perhaps, I had better content myself now with calling your attention to the fact that the statistical aspect of the matter is in a different position now. The assertions that a rise in prices increased the real wages of the worker were so consonant with the common opinion of people who had not specially studied the matter, that it was accepted almost as an axiom; but, within the last ten years, the statistics of wages have been carried so far in certain countries, and especially in England and America, that we are able to bring it to the test. I have accumulated a great number of facts, but nearly everything I have accumulated is implied in this table. It is copied from the article by Mr Bowley in the *Economic Journal* for last December. It is the result of work that has been going on for a number of years, and seems to me to be practically decisive. It

[1] *Official Papers*, pp. 284–288.

collects the average wages in England from the year 1844 to the year 1891, and then calculates what purchasing power the wages would give at the different times, and it shows that the rise of real wages after 1873 when prices were falling was greater than before 1873 when prices were rising.

Here follows a table from Prof. Bowley's article in this *Journal* for December 1898. Marshall's final conclusion was crystallised in a passage in the *Principles* (Book VI, ch. VIII, §6):

> '[When prices rise the employer] will therefore be more able and more willing to pay the high wages; and wages will tend upwards. But experience shows that (whether they are governed by sliding scales or not) they seldom rise as much in proportion as prices; and therefore they do not rise nearly as much in proportion as profits.'

Although Marshall's evidence before the Indian Currency Committee was given in 1899, Prof. Bowley's statistics on which he was relying do not relate effectively to a date later than 1891 (or 1893 at latest). It is clear, I think, that Marshall's generalisation was based on experience from 1880 to 1886 which did in fact bear it out. If we divide the years from 1880 to 1914 into successive periods of recovery and depression, the broad result, allowing for trend, appears to be as follows:

		Real wages
1880–1884	Recovery	Falling
1884–1886	Depression	Rising
1886–1890	Recovery	Rising
1890–1896	Depression	Falling
1896–1899	Recovery	Rising
1899–1905	Depression	Falling
1905–1907	Recovery	Rising
1907–1910	Depression	Falling
1910–1914	Recovery	Rising

According to this, Marshall's generalisation holds for the periods from 1880 to 1884 and from 1884 to 1886, but for no subsequent periods.[1] It seems that we have been living all these years on a generalisation which held good, by exception, in the years 1880–86, which was the formative period in Marshall's thought in this matter, but has never once held good in the fifty years since he crystallised it! For Marshall's view mainly prevailed, and Foxwell's contrary opinion was discarded as the heresy of an

---

[1] I compiled this table, as a check, independently of Mr Dunlop's table, *loc. cit.*, p. 419. But it only serves to confirm his more accurate version. According to him, trend eliminated, real wages fell 3 per cent in the recovery culminating in 1883 or 1884 and rose 2·7 per cent in the depression from 1884 to 1886.

inflationist. It is to be observed that Marshall offered his generalisation merely as an observed statistical fact, and, beyond explaining it as probably due to wages being stickier than prices, he did not attempt to support it by *à priori* reasoning. The fact that it has survived as a dogma confidently accepted by my generation must be explained, I think, by the more theoretical support which it has subsequently received.

To my statement that Marshall's generalisation has remained uncorrected until recently there is, however, an important exception. In his *Industrial Fluctuations*, published in 1927, Professor Pigou pointed out (p. 217) that 'the upper halves of trade cycles have, on the whole, been associated with higher rates of real wages than the lower halves,' and he printed in support of this a large scale chart for the period from 1850 to 1910. Subsequently, however, he seems to have reverted to the Marshallian tradition, and in his *Theory of Unemployment*, published in 1933, he writes (p. 296):

> In general, the translation of inertia from real wage-rates to money wage-rates causes real rates to move in a manner not compensatory, but complementary, to movements in the real demand function. Real wage-rates not merely fail to fall when the real demand for labour is falling, but actually rise; and, in like manner, when the real demand for labour is expanding, real wage-rates fall.

About that time M. Rueff had attracted much attention by the publication of statistics which purported to show that a rise in real wages tended to go with an increase in unemployment, Prof. Pigou points out that these statistics are vitiated by the fact that M. Rueff divided money wages by the wholesale index instead of by the cost-of-living index, and he does not agree with M. Rueff that the observed rise in real wages was the main *cause* of the increased unemployment with which it was associated. But he concludes, nevertheless (p. 300), on a balance of considerations, that 'there can be little doubt that in modern industrial communities this latter tendency (i.e. for shifts in real demand to be associated with shifts in the *opposite* sense in the rate of real wages for which work people stipulate) is predominant'.

Like Marshall, Prof. Pigou based his conclusion primarily on the stickiness of money wages relatively to prices. But my own readiness to accept the prevailing generalisation, at the time when I was writing my *General Theory*, was much influenced by an *à priori* argument, which had recently won wide acceptance, to be found in Mr R. F. Kahn's article on 'The Relation of Home Investment to Employment,' published in the *Economic Journal* for June 1931.[1] The supposed empirical fact, that in the short period real wages tend to move in the opposite direction to the level of output, appeared, that is to say, to be in conformity with the more fundamental generalisations that industry is subject to increasing marginal cost in the short period, that for a closed[2] system as a whole marginal cost in the short period is substantially the same thing as marginal wage cost, and that in competitive conditions prices are governed by marginal cost; all this being subject, of course, to various qualifications in particular cases, but remaining a reliable generalisation by and large.

I now recognise that the conclusion is too simple, and does not allow sufficiently for the complexity of the facts. But I still hold to the main structure of the argument, and believe that it needs to be amended rather than discarded. That I was an easy victim of the traditional conclusion because it fitted my theory is the opposite of the truth. For my own theory this conclusion was inconvenient, since it had a tendency to offset the influence of the main forces which I was discussing and made it necessary for me to introduce qualifications, which I need not have troubled with if I could have adopted the contrary generalisation favoured by Foxwell, Mr Dunlop and Mr Tarshis. In particular, the traditional conclusion played an important part, it will be remembered, in the discussions, some ten years ago, as to the effect of expansionist policies on employment, at a time when I had not developed my own argument in as complete a form as I did subsequently. I was already arguing at that time that the good effect of an expansionist investment policy on employment, the fact of which no one denied, was due to the stimulant which it gave to effective demand. Prof. Pigou, on the other hand, and many other economists explained the observed result by the reduction in real wages covertly effected by the rise in prices which ensued on the increase in effective demand. It was held that

---

[1] *Passim*; see particularly pp. 178, 182. It was Mr Kahn who first attacked the relation of the general level of prices to wages in the same way as that in which that of particular prices has always been handled, namely as a problem of demand and supply in the short period rather than as a result to be derived from monetary factors.

[2] The qualifications required, if the system is not closed, are dealt with below.

public investment policies (and also an improvement in the trade balance through tariffs) produced their effect by deceiving, so to speak, the working classes into accepting a lower real wage, effecting by this means the same favourable influence on employment which, according to these economists, would have resulted from a more direct attack on real wages (e.g. by reducing money wages whilst enforcing a credit policy calculated to leave prices unchanged). If the falling tendency of real wages in periods of rising demand is denied, this alternative explanation must, of course, fall to the ground. Since I shared at the time the prevailing belief as to the facts, I was not in a position to make this denial. If, however, it proves right to adopt the contrary generalisation, it would be possible to simplify considerably the more complicated version of my fundamental explanation which I have expounded in my *General Theory*.[1] My practical conclusions would have, in that case, *à fortiori* force. If we can advance farther on the road towards full employment than I had previously supposed without seriously affecting real hourly wages or the rate of profits per unit of output, the warnings of the anti-expansionists need cause us less anxiety.

Nevertheless, we should, I submit, hesitate somewhat and carry our inquiries further before we discard too much of our former conclusions which, subject to the right qualifications, have *à priori* support and have survived for many years the scrutiny of experience and common sense. I offer,[2] therefore, for further statistical investigation an analysis of the elements of the problem with a view to discovering at what points the weaknesses of the former argument emerge. There are five heads which deserve separate consideration.

I

First of all, are the statistics on which Mr Dunlop and Mr Tarshis are relying sufficiently accurate and sufficiently uniform in their indications to form the basis of a reliable induction?

For example, in so recent a compilation as the League of Nations *World Economic Survey 1937–38*, prepared by Mr J. E. Meade, the traditional conclusion receives support, not on *à priori*

---

[1] Particularly in Chapter 2, which is the portion of my book which most needs to be revised.

[2] In amplification of Mr Dunlop's useful summary at the end of his article (*loc. cit.*, pp. 431–3).

grounds, but on the basis of the most recently available statistics. I quote the following from pp. 54–55:

> During the great depression after 1929, the demand for goods and services diminished, and in consequence the price of commodities fell rapidly. In most countries, as can be seen from the graph on p. 52, hourly money wages were reduced as the demand for labour fell; but in every case there was a greater fall in prices, so that hourly real wages rose...[It is then explained that the same was not true of weekly wages.]...Since the recovery, the opposite movements may be observed. In most countries, increased demand for goods and services has caused commodity prices to rise more rapidly than hourly money wages, and the hourly real wage has fallen...In the United States[1] and France,[2] however, the rise in money wages was so rapid between 1936 and 1937 that the hourly real wage continued to rise...When real hourly wages are raised—i.e. when the margin between commodity prices and the money-wage cost becomes less favourable—employers are likely to diminish the amount of employment which they offer to labour. While there were, no doubt, other influences affecting the demand for labour, the importance of this factor is well illustrated by the graph on p. 53. In the case of all the countries represented for which information is available, the fall in commodity prices between 1929 and 1932 caused a rise in the hourly real wage, and this was accompanied by a diminution in employment...(it is shown that on the recovery there has been a greater variety of experience)...

This authoritative study having international scope indicates that the new generalisations must be accepted with reserve. In any case Mr Tarshis's scatter diagram printed below [in the *Economic Journal*, March 1939] (p. 150), whilst it shows a definite preponderance in the south-west and north-east compartments and a high coefficient of association, includes a considerable number of divergent cases, and the absolute range of most of the scatter is extremely small, with a marked clustering in the neighbourhood of the zero line for changes in real wages; and much the same is true of Mr Dunlop's results. The great majority of Mr Tarshis's observations relate to changes of less than 1·5 per cent. In the introduction to his *Wages and Income in the United Kingdom since 1860*, Prof. Bowley indicates that this is probably less than the margin of error for statistics of this kind. This general conclusion

[1] [Probably as a result of the New Deal.]
[2] Explained as being due to the forty-hour week.

is reinforced by the fact that it is *hourly* wages which are relevant in the present context, for which accurate statistics are not available.[1] Moreover, in the *post-scriptum* to his note, Mr Tarshis explains that whilst real wages tend to move in the same direction as money wages, they move in the opposite direction, though only slightly, to the level of output as measured by man-hours of employment; from which it appears that Mr Tarshis's final result is in conformity with my original assumption, which is, of course, concerned with hourly wages. It seems possible, therefore, taking account of Mr Meade's results, that I may not, after all, have been seriously wrong.

Furthermore, for reasons given below, it is important to separate the observations according as the absolute level of employment is distinctly good or only mediocre. It may be that we can analyse our results so as to give two distinct generalisations according to the absolute level reached by employment. If, at the present stage of the inquiry, we are to make any single statistical generalisation, I should prefer one to the effect that, for fluctuations within the range which has been usual in the periods investigated which seldom approach conditions of full employment, short-period changes in real wages are usually so small compared with the changes in other factors that we shall not often go far wrong if we treat real wages as substantially constant in the short period (a very helpful simplification if it is justified). The conclusion, that changes in real wages are not usually an important factor in short-period fluctuations until the point of full employment is approaching, is one which has been already reached by Dr Kalecki on the basis of his own investigations.[2]

II

It may be that we have under-estimated the quantitative effect of a factor of which we have always been aware. Our argument assumed that, broadly speaking, labour is remunerated in terms of its own composite product, or at least that the price of wage-goods moves in the same way as the price of output as a whole. But no one has supposed that this was strictly the case or was better than an approximation; and it may be that the

---

[1] It is possible that Mr Meade has been more successful than Mr Dunlop in using hourly wages, and that this explains some discrepancies in their conclusions.

[2] 'The Determinants of Distribution of the National Income' *Econometrica*, April 1938, p. 102, now reprinted in his *Essays in the Theory of Economic Fluctuations*.

proportion of wage-goods, which are not the current product of the labour in question and the prices of which are not governed by the marginal cost of such product, is so great as to interfere with the reliability of our approximation. House-rent and goods imported on changing terms of trade are leading examples of this factor. If in the short period rents are constant and the terms of trade tend to improve when money wages rise and to deteriorate when money wages fall, our conclusion will be upset in practice in spite of the rest of our premises holding good.

In the case of this country one has been in the habit of supposing that these two factors have in fact tended to offset one another, though the opposite might be the case in the raw-material countries. For whereas rents, being largely fixed, rise and fall less than money wages, the price of imported food-stuffs tends to rise more than money wages in periods of activity and to fall more in periods of depression. At any rate both Mr Dunlop and Mr Tarshis claim to show that fluctuations in the terms of trade (terms of foreign trade in Mr Dunlop's British inquiry and terms of trade between industry and agriculture in Mr Tarshis's American inquiry) are not sufficient to affect the general tendency of their results, though they clearly modify them quantitatively to a considerable extent.[1] Nevertheless, the effect of expenditure on items such as rent, gas, electricity, water, transport, etc., of which the prices do not change materially in the short period, needs to be separately calculated before we can be clear. If it should emerge that it is this factor which explains the results, the rest of our fundamental generalisations would remain undisturbed. It is important, therefore, if we are to understand the situation, that the statisticians should endeavour to calculate wages in terms of the actual product of the labour in question.

III

Has the identification of marginal cost with marginal wage cost introduced a relevant error? In my *General Theory of Employment*, chapter 6 (Appendix), I have argued that this identification is dangerous in that it ignores a factor which I have called 'marginal user cost'. It is unlikely, however, that this can help us in the present context. For marginal user cost is likely to increase when output is increasing, so that this factor

[1] Cf. Dunlop, *loc. cit.*, p. 417.

would work in the opposite direction from that required to explain our present problem, and would be an additional reason for expecting prices to rise more than wages. Indeed, one would, on general grounds, expect marginal total cost to increase more, and not less, than marginal wage cost.

IV

Is it the assumption of increasing marginal real cost in the short period which we ought to suspect? Mr Tarshis finds part of the explanation here; and Dr Kalecki is inclined to infer approximately constant marginal real cost.[1] But there is an important distinction which we have to make. We should all agree that if we start from a level of output very greatly below capacity, so that even the most efficient plant and labour are only partially employed, marginal real cost may be expected to decline with increasing output, or, at the worst, remain constant. But a point must surely come, long before plant and labour are fully employed, when less efficient plant and labour have to be brought into commission or the efficient organisation employed beyond the optimum degree of intensiveness. Even if one concedes that the course of the short-period marginal cost curve is downwards in its early reaches, Mr Kahn's assumption that it eventually turns upwards is, on general common-sense grounds, surely beyond reasonable question; and that this happens, moreover, on a part of the curve which is highly relevant for practical purposes. Certainly it would require more convincing evidence than yet exists to persuade me to give up this presumption.

Nevertheless, it is of great practical importance that the statisticians should endeavour to determine at what level of employment and output the short-period marginal-cost curve for the composite product as a whole begins to turn upward and how sharply it rises after the turning-point has been reached. This knowledge is essential for the interpretation of the trade cycle. It is for this reason that I suggested above that the observations of the relative movement of real and money wages should be separately classified according to the average level of employment which had been reached.

It may prove, indeed, at any rate in the case of statistics relating to recent years that the level of employment has been

---

[1] *Loc. cit.*

preponderantly so low that we have been living more often than not on the reaches of the curve before the critical point of upturn has been attained. It should be noticed that Mr Tarshis's American figures relate only to the period from 1932 to 1938, during the whole of which period there has been such intense unemployment in the United States, both of labour and of plant, that it would be quite plausible to suppose that the critical point of the marginal cost curve had never been reached. If this has been the case, it is important that we should know it. But such an experience must not mislead us into supposing that this must necessarily be the case, or into forgetting the sharply different theory which becomes applicable after the turning-point has been reached.

If, indeed, the shape of the marginal-cost curve proves to be such that we tend to be living, with conditions as they are at present, more often to the left than to the right of its critical point, the practical case for a planned expansionist policy is considerably reinforced; for many *caveats* to which we must attend after this point has been reached can be, in that case, frequently neglected. In taking it as my general assumption that we are often on the right of the critical point, I have been taking the case in which the practical policy which I have advocated needs the most careful handling. In particular the warnings given, quite rightly, by Mr D. H. Robertson of the dangers which may arise when we encourage or allow the activity of the system to advance too rapidly along the upward slopes of the marginal-cost curve towards the goal of full employment, can be more often neglected, for the time being at least, when the assumption which I have previously admitted as normal and reasonable is abandoned.

v

There remains the question whether the mistake lies in the approximate identification of marginal cost with price, or rather in the assumption that for output as a whole they bear a more or less proportionate relationship to one another irrespective of the intensity of output. For it may be the case that the practical workings of the laws of imperfect competition in the modern quasi-competitive system are such that, when output increases and money wages rise, prices rise less than in proportion to the increase in marginal money cost. It is scarcely likely, perhaps, that the narrowing gap could be sufficient to prevent a decline in real wages in a phase in which marginal real cost was increasing rapidly. But it might be sufficient to offset the effect on real wages of a modest rise in marginal real cost, and even to dominate the situation in the event of the marginal real cost curve proving to be almost horizontal over a substantial portion of its relevant length.

It is evidently possible that some such factor should exist. It might be, in a sense, merely an extension of the stickiness of prices of which we have already taken account in II above. Apart from those prices which are virtually constant in the short period, there are obviously many others which are, for various reasons, more or less sticky. But this factor would be particularly likely to emerge when output increases, in so far as producers are influenced in their practical price policies and in their exploitation of the opportunities given them by the imperfections of competition, by their long-period average cost, and are less attentive than economists to their short-period marginal cost. Indeed, it is rare for anyone but an economist to suppose that price is predominantly governed by marginal cost. Most business men are surprised by the suggestion that it is a close calculation of short-period marginal cost or of marginal revenue which should dominate their price policies. They maintain that such a policy would rapidly land in bankruptcy anyone who practised it. And if it is true that they are producing more often than not on a scale at which marginal cost is falling with an increase in output, they would clearly be right; for it would be only on rare occasions that they would be collecting anything whatever towards their overhead. It is, beyond doubt, the practical assumption of the producer that his price policy ought to be influenced by the fact that he is normally operating subject to decreasing average cost, even if in the short-period his marginal cost is rising. His effort is to maintain prices when output falls and, when output increases, he may raise them by less than the full amount required to offset higher costs including higher wages. He would admit that this, regarded by him as the reasonable, prudent and far-sighted policy, goes by the board when, at the height of the boom, he is overwhelmed by more orders than he can supply; but even so he is filled with foreboding as to the ultimate consequences of his being forced so far from the right and reasonable policy of fixing his prices by reference to his long-period overhead as well as his current costs. Rightly ordered competition consists, in his opinion, in a proper pressure to secure an adjustment of prices to changes in long-period average cost; and the suggestion that he is becoming a dangerous and anti-social monopolist whenever, by open or tacit agreement with his competitors, he endeavours to prevent prices from following short-period marginal cost, however much this may fall away from long-period average cost, strikes him as disastrous. (It is the failure of the latest phase of the New Deal in the United States, in contrast to the earliest phase, of which the opposite is true, to distinguish between price agreements for maintaining prices in right relation to average long-period cost and those which aim at obtaining a monopolistic profit in excess of average long-period cost which strikes him as particularly unfair.)

Thus, since it is the avowed policy of industrialists to be content with a smaller gross profit per unit of output when output increases than when it declines, it is not unlikely that this policy may be, at least partially, operative. It would be of great interest if the statisticians could show in detail in what way gross profit per unit of output changes in different industries with a changing ratio between actual and capacity output. Such an investigation should distinguish, if possible, between the effect of increasing output on unit-profit and that of higher costs in the shape of higher money wages and other expenses. If it should appear that increasing output as such has a tendency to decrease unit-profit, it would follow that the policy suggested above is actual as well as professed. If, however, the decline in unit-profit appears to be mainly the result of a tendency of prices to offset higher costs incompletely, irrespective of changes in the level of output, then we have merely an example of the stickiness of prices arising out of the imperfection of competition intrinsic to the market conditions. Unfortunately it is often difficult or impossible to distinguish clearly between the effects of the two influences, since higher money costs and increasing output will generally go together.

A well-known statistical phenomenon which ought to have put me on my guard confirms the probability of constant or diminishing, rather than increasing, profit per unit of output when output increases. I mean the stability of the proportion of the national dividend accruing to labour, irrespective apparently of the level of output as a whole and of the phase of the trade cycle. This is one of the most surprising, yet best-established, facts in the whole range of economic statistics, both for Great Britain and for the United States. The following figures summarise briefly what are, I believe, the undisputed facts:[1]

RELATIVE SHARE OF MANUAL LABOUR IN THE NATIONAL INCOME OF GREAT BRITAIN[2]

1911	40·7	1924	43·0	1928	43·0	1932	43·0
		1925	40·8	1929	42·4	1933	42·7
		1926	42·0	1930	41·1	1934	42·0
		1927	43·0	1931	43·7	1935	41·8

RELATIVE SHARE OF MANUAL LABOUR IN THE NATIONAL INCOME OF U.S.A.[3]

1919	34·9	1923	39·3	1927	37·0	1931	34·9
1920	37·4	1924	37·6	1928	35·8	1932	36·0
1921	35·0	1925	37·1	1929	36·1	1933	37·2
1922	37·0	1926	36·7	1930	35·0	1934	35·8

The fluctuations in these figures from year to year appear to be of a random character, and certainly give no significant indications of any tendency to move against labour in years of increasing output. It is the stability of the ratio for each country which is chiefly remarkable, and this appears to be a long-run, and not merely a short-period, phenomenon.[4] Moreover, it would be interesting to discover whether the difference between the British and the American ratio is due to a discrepancy in the basis of reckoning adopted in the two sets of statistics or to a significant difference in the degrees of monopoly prevalent in the two countries or to technical conditions.

In any case, these facts do not support the recently prevailing assumptions as to the relative movements of real wages and output, and are inconsistent with the idea of there being any marked tendency to increasing unit-profit with increasing output. Indeed, even in the light of the above considerations, the result remains a bit of a miracle. For even if price policies are such as

[1] The British figures are based on Mr Colin Clark's *National Income and Outlay*, and the American figures on Dr King's *The National Income and its Purchasing Power, 1909–1928*, and Dr Kuznets' *National Income and Capital Formation, 1919–1935*. But in both cases I have used the slightly adjusted version of the figures prepared by Dr Kalecki and given by him in his *Essays in the Theory of Economic Fluctuations*, pp. 16, 17.

[2] Shop assistants excluded.

[3] Shop assistants included.

[4] Dr Bowley has given a figure of 41·4 for Great Britain in 1880. Dr Kalecki tells me that, if this was adjusted so as to be comparable with the figures given above, it would be about 42·7—which would show an extraordinary stability for the ratio over a period of no less than fifty-five years during which almost everything else changed out of knowledge.

to cause unit-profit to decrease in the same circumstances as those in which marginal real cost is increasing, why should the two quantities be so related that, regardless of other conditions, the movement of the one *almost exactly* offsets the movement of the other? I recently offered the problem of explaining this ἀπορία, as Edgeworth would have called it, to the research students at Cambridge. The only solution was offered by Dr Kalecki in the brilliant article which has been published in *Econometrica*.[1] Dr Kalecki here employs a highly original technique of analysis into the distributional problem between the factors of production in conditions of imperfect competition, which may prove to be an important piece of pioneer work. But the main upshot is what I have indicated above, and Dr Kalecki makes, to the best of my understanding, no definite progress towards explaining why, when there is a change in the ratio of actual to capacity output, the corresponding changes in the degree of the imperfection of competition should *so exactly* offset other changes. Nor does he explain why the distribution of the product between capital and labour should be stable in the long run, beyond suggestion that changes of one kind always just serve to offset changes of another; yet it is very surprising that on balance there should have been a constant degree of monopoly over the last twenty years or longer. His own explanation is based on the assumptions that marginal real costs are constant, that the degree of the imperfection of the market changes in the opposite direction to output, but that this change is precisely offset by the fact that the prices of basic raw materials (purchased by the system from outside) relatively to money wages increase and decrease with output. Yet there is no obvious reason why these changes should so nearly offset one another; and it would seem safer not to assume that marginal real costs are constant, but to conclude that in actual fact, when output changes, the change in the degree of the imperfection of the market is such as to offset the combined effect of changes in marginal costs and of changes in the prices of materials bought from outside the system relatively to money wages. It may be noticed that Dr Kalecki's argument assumes the existence of an opposite change in the degree of the imperfection of competition (or in the degree in which producers take advantage of it) when output increases from that expected by Mr R. F.

[1] April 1938, 'The Determinants of Distribution of the National Income', and now reprinted in his book referred to above.

Harrod in his study on *The Trade Cycle*. There Mr Harrod expects an increase; here constancy or a decrease seems to be indicated. Since Mr Harrod gives grounds for his conclusions which are *prima facie* plausible, this is a further reason for an attempt to put the issue to a more decisive statistical test.[1]

To state the case more exactly, we have five factors which fluctuate in the short period with the level of output:

(1) The price of wage-goods relatively to the price of the product;

(2) The price of goods bought from outside the system relatively to money wages;

(3) The marginal wage cost;

(4) The marginal user cost (I attach importance to including this factor because it helps to bridge the discontinuity between an increase of output up to short-period capacity and an increase of output involving an increase beyond the capacity assumed in short-period conditions); and

(5) The degree of the imperfection of competition.

And it appears that, for reasons which are not yet clear, these factors taken in conjunction have no significant influence on the distribution between labour and capital of the income resulting from the output. Whatever a more complete inquiry into the problem may bring forth, it is evident that Mr Dunlop, Mr Tarshis and Dr Kalecki have given us much to think about, and have seriously shaken the fundamental assumptions on which the short-period theory of distribution has been based hitherto;— it seems that for practical purposes a different set of simplifications from those adopted hitherto are preferable. Meanwhile I am comforted by the fact that their conclusions tend to confirm the idea that the causes of short-period fluctuation are to be found in changes in the demand for labour, and not in changes in its real-supply price; though I complain a little that I in particular should be criticised for conceding a little to the other view by admitting that, when the changes in effective demand to which I myself attach importance have brought about a change in the level of output, the real-supply price for labour would in fact change in the direction assumed by the theory I am opposing—as if I was the

---

[1] Dr Kalecki's conclusion is in conformity with Prof. Pigou's argument in *Industrial Fluctuations*, Book I, chap. XVIII, where reasons are given for expecting more imperfection of competition in depressions.

first to have entertained the fifty-year-old generalisation that, trend eliminated, increasing output is usually associated with a falling real wage.

I urge, nevertheless, that we should not be too hasty in our revisions, and that further statistical enquiry is necessary before we have a firm foundation of fact on which to reconstruct our theory of the short period. In particular we need to know:

(i) How the real hourly wage changes in the short period, not merely in relation to the money wage, but in relation to the percentage which actual output bears to capacity output;

(ii) How the purchasing power of the industrial money wage in terms of its own product changes when output changes; and

(iii) How gross profit per unit of output changes (*a*) when money costs change, and (*b*) when output changes.

J. M. KEYNES

INDEX